Brain Sense

The Science of the Senses and How We Process the World Around Us

Faith Hickman Brynie

AMACOM

American Management Association
New York • Atlanta • Brussels • Chicago • Mexico City
San Francisco • Shanghai • Tokyo • Toronto • Washington, D.C.

Special discounts on bulk quantities of AMACOM books are available to corporations, professional associations, and other organizations. For details, contact Special Sales Department, AMACOM, a division of American Management Association, 1601 Broadway, New York, NY 10019.
Tel: 800-250-5308. Fax: 518-891-2372.
E-mail: specialsls@amanet.org
Website: www.amacombooks.org/go/specialsales
To view all AMACOM titles go to: www.amacombooks.org

Library of Congress Cataloging-in-Publication Data
Brynie, Faith Hickman
Brain sense : the science of the senses and how we process the world around us / Faith Hickman Brynie.
p. cm.
Includes bibliographical references and index.
ISBN-13: 978-0-8144-1324-1
ISBN-10: 0-8144-1324-2
1. Senses and sensation. I. Title.
BF233.B97 2010
152.1—dc22 2009017998

Illustrations by Susan Gilbert, CMI.

Printing number
10 9 8 7 6 5 4 3 2 1

I have been told that my need to understand things is a disease. This book is dedicated to all those who are similarly afflicted.

Our brains are good for getting us around and mating successfully, and even for doing some serious physics, but they go blank when they try to understand how they produce the awareness that is our prized essence. The consolation is that we shall always be of intense interest to ourselves, long after quantum theory has become old hat.

—Colin McGinn
"An Unbridgeable Gulf"
2007

CONTENTS

PART THREE TASTE

PART FOUR VISION

PART FIVE HEARING

PART SIX BEYOND THE BIG FIVE

Preface: Falling in Love with Science

I don't recall when I first fell in love with science, but I remember the day when I said, "'Til death do us part." I was counting raspberry bushes. They grew wild around the abandoned strip mines of Appalachia. As an ecology student at West Virginia University in Morgantown, I clambering around an old mine's precarious slopes with twenty other eager undergraduates. We shot line transects and counted the bushes, orienting our test sites by the compass, while measuring roped-off segments ten-meters square for careful counting and mapping. The day was hot and sticky. The prickly bushes tore our clothes and gouged our flesh. Black coal dust clogged our lungs. Sunburned and sweaty, we learned that wrestling truth from reality was difficult . . . and fun!

Field science infatuated me that day, but my pledge of lifelong devotion to the scientific process came a few days later, when we pooled data from several teams. We made graphs of numbers of raspberry bushes, north and east, upslope and down. The graphs sang to me. Their meaning popped off the page and danced around my desk. In axes, points, clusters, and lines, the numbers of raspberry bushes revealed the history of the mine. In the days long before ecology became a household word, those bars, dots, lines, and curves mirrored the fifty-year history of the mine, disclosing how the site had been worked, when it had been abandoned, where the acid mine drainage had polluted most, and how nature had attempted—with wild raspberries—to bandage the land so it could heal from within. The data painted a picture more beautiful to me than any art museum masterpiece.

From that day on, I never questioned my choice of a career. It was science for me, in some form, and I've tried quite a few. In the bacteriology labs at West

Virginia University, I grew on agar culture plates the acidophilic microorganisms that can survive in acid mine drainage when nothing else can. At Oak Ridge National Laboratories, I zapped protozoans with ultraviolet light to see if they would die. They did. At West Virginia University Medical Center, I worked under a grant from the U.S. Army, screening the effects of possible new malarial drugs on blood samples run through a labyrinth of tubes, reaction vessels, and colorimeters. (Computers do all that now.) In Colorado, I mixed reagents for thyroid tests in a medical lab. I stuck thermometers down prairie dog burrows in Rocky Mountain National Park, and I set up anemometers in mountain meadows. Then I got into science teaching and educational reform because I wanted every young person to fall in love with science just as I had. Writing curriculum and designing educational materials led me into topics ranging from medical genetics to land use planning.

In the early 1990s, I read a sentence that riveted my attention. The neuroscientist Richard Restak wrote, "The [human] brain is the only organ in the known universe that seeks to understand itself."[1] That sentence stopped me dead in my tracks. Here was the science of science itself—an organ of tissue and blood attempting to understand its own functioning. To my mind, that was bigger than malaria drugs, bigger than burrowing animals, bigger even than my raspberry bushes. I couldn't wait to find out what neuroscientists were doing with their brains as they attempted to comprehend . . . their own brains! So I started digging though the scientific literature and eventually I wrote a book called *101 Questions Your Brain Has Asked About Itself But Couldn't Answer . . . Until Now.* The book was moderately successful, garnering a "Best Book" honor from the American Association for the Advancement of Science in 1999. The book is now in its second edition and continuing to make a contribution, I hope.

For me, however, one hundred and one questions formed only the tip of the iceberg. The more questions I asked and the more I read and learned, the more I wanted to know. I became fascinated with research on how the brain and senses interact—thus this book's title, *Brain Sense.* "Brain sense" is a field that many tillers are plowing: the neurologists who study the interaction of peripheral and central nervous system; the brain mappers who chart the regions of the brain that handle specialized functions and the nerve networks that connect those regions; the biochemists who probe the molecular receptors that initiate the impulses of the chemical senses; the biophysicists who explore how light and sound waves can sometimes translate themselves into sensory perceptions; the physicians and surgeons who seek to treat the maladies that befall the senses and the brain; the engineers and biomechanicists who try to understand how perception works so they can construct devices ranging from prosthetics to virtual reality simulators; the cognitive psychologists who want to understand how we learn; the behavioral psychologists who hope to explain why we do the things we do; and the clinical psychologists who strive to cure the intellectual, social, and emotional sequelae of sensory perception gone awry.

The organization of this book follows traditional lines. There's a part on each of the five major senses: touch, smell, taste, vision, and hearing. The book begins with touch because, in my opinion, it's the least appreciated of all the senses and perhaps the most vital. We explore the chemical senses next, because taste and smell work in similar ways. Next come sight and sound, the senses that we rely on the most and can trust the least.

In the individual chapters about each sense, I've tried to include some stories about how ordinary people live when their sensory capacities are diminished or lost. I've also included as much of the latest in scientific research as I could jam onto each page. There is so much going on in brain-sense research, I can scarcely scratch the surface here, but I hope that the studies I've chosen for inclusion impart some sense of the endless fascination of that work.

Throughout these chapters, you'll notice two recurring themes. The first is brain plasticity. Plasticity means that the brain changes throughout life. Once, we thought of the adult brain as fixed and unchanging (except for the deterioration that comes with age), but research in the last two decades has shattered that notion. The brain is constantly rewiring itself, reshaping its own structure, recruiting new circuits to perform its work when old ones are damaged or lost. It even re-creates memories each time it retrieves them. The implications of brain plasticity for understanding our senses, our consciousness, and the essence of what it means to be human are nothing short of staggering.

The second theme is what I've come to think of as a negotiable reality. We believe in our senses, and we trust that they are providing us with objective, complete, and accurate data about the world around us. We are wrong. Our brains construct our reality, molding every input to what we expect, what we imagine, what we wish for. Our brains have minds of their own. They shape light waves into expected images, sound waves into patterns ranging from noise to music. Our sense of touch is malleable, variable, and refinable. We taste and smell what we believe we will taste and smell. In precisely the same environment, you and I will neither see, hear, taste, touch, nor smell the same things—nor will we draw the same conclusions about the information our senses have collected. Our personal worlds are constructions built by our brains using the raw materials of the senses—raw materials that are greatly modified during the construction process.

That idea of a negotiable reality is the reason for the last part of this book, "Beyond the Big Five," which looks briefly at some of the other "senses" or sensory experiences that don't fit with the big five but are too intriguing to ignore, such as the mingling of the senses known as synesthesia, the experience of déjà vu, the phantom sensations and phantom pain often experienced by people who have amputations, the possibilities and probabilities of extraordinary sensory perception, and the brain's sense of time kept by a body clock far more precise than most of us realize.

I hope that people who know a lot about the brain and the senses will read this book. I hope that people who know very little about those topics will read it, too. For those readers who are interested in the brain and the senses but don't know much about the brain's structure and function, I've included an appendix at the end of the book, "The Brain and the Nervous System—A Primer," which provides a short course in neuroscience and explains many of the terms and concepts used in this book. It also includes diagrams showing the locations of many of the brain regions discussed in the book. Before long, you'll know your occipital lobe from your parietal, and you'll be well on your way to comprehending your own "brain sense."

From beginning to end, this book is many things. It's part memoir because it's my opportunity to reminisce about some things I've learned from science and from life. It's part investigative reporting because I've delved into the work of some cutting-edge researchers who are designing clever experiments to gain answers to questions that we didn't even know how to ask a decade ago. It's part biography because I want you to know—as I have come to know—what real scientists are like as they work in real labs on real questions that have never before been answered. It's part textbook because basic knowledge about how our senses work is important to everyone. It's part history because we can't appreciate where we're going if we don't value where we've been. It's part newspaper because it contains some of the late-breaking stories that are making headlines on a daily basis. It's part travel journal because I invite you to fly with me as I visit working neuroscientists in Washington, Minnesota, Michigan, and Massachusetts. It's part personality profiles because the scientists I met and talked with are intriguing people, doing interesting work and living full and satisfying lives. I want readers of this book to see scientists as I have seen them—not as geeky weirdos in lab coats, but as warm, humorous, engaging human beings who thrive on the search for knowledge as they toil among the "raspberry bushes" of their chosen research specialty.

Most of all, this book is a tribute to courage and to some of the wonderful people who shared their stories with me: the tour guide who faced death as her ship sank in Antarctica; the hairdresser who lost her sense of smell to brain injury; the woman who had a mastectomy but continues to feel her breasts; the young poet, born deaf, who had a cochlear implant; the synesthete who sees letters in colors; the electronic genius who engineers phone lines and ignores his tinnitus; the teenager who was born without the sense of touch and the mother who has loved him totally and unconditionally through all he has achieved.

Finally, this book is a love letter to science and scientists. I've been wedded to science all my life, and my fascination with the questions, methods, and inferences of scientific research has never diminished. Science isn't the only way to see, to know, to understand, but it's the one that won my heart. Come meet the love of my life.

Acknowledgments

Many institutions and individuals provided valuable information in the research phases of this book. I've drawn on the work of hundreds of scientists—too many to list—whom I admire deeply and thank wholeheartedly. I exchanged emails with dozens of them. I enjoyed speaking with some of them. I traveled to visit a few of them. I treasure them all. Whatever is right in this book is to their credit. Whatever is wrong can be attributed to my fallibility.

Special thanks go to those scientists who allowed me to visit their laboratories and who took time from their busy academic schedules to meet with me and teach me those things that are best learned human to human, face to face. Special appreciation to Scott Murray, University of Washington; Shawn Green, University of Minnesota; Wayne Aldridge and Steve Mahler, University of Michigan; Susan Brown, University of Michigan; Daniel Goble (by webcam), Katholieke Universiteit Leuven, Heverlee, Belgium; Lucia Vaina, Boston University; Mandayam Srinivasan, Massachusetts Institute of Technology; and Gottfried Schlaug and Andrea Norton, Department of Neurology, Beth Israel Deaconess Medical Center in Boston. Also special thanks to some I've never met but hope to some day: phantom limb expert Emma Duerden, psychotherapist Nanette Mongelluzzo, and language researcher Ghislaine Dehaene-Lambertz.

I appreciate the efforts of my agent, Grace Freedson, in making this book possible. Thanks to Amacom editor Barry Richardson for his enthusiasm and encouragement. Thanks also to production editor Michael Sivilli and editorial wizard Barbara Chernow for all their hard work.

I am most grateful to those individuals who selflessly shared their personal stories, including Chuck Aoki, Dave Chappelle, Theresa Dunlop, Marv Gisser, Karyn Maier, Jennifer Nelson, Andrea Salas, and Stephen Michael Verigood. And to Kim McKee, the poet, my gratitude; I wish I could have used one of her poems in every chapter.

The greatest gratitude is earned by those who tolerate on a daily basis the rigors of a process that bears little reward beyond itself. Ann, you are always in my heart and thoughts, and now Rob and Katherine join you there. As for you, Lloyd, you make sense of every day for me, and yours is the kindest heart and the best brain of all.

PART ONE
TOUCH

Children are born true scientists. They spontaneously experiment and experience and reexperience again. They . . . touch-test for hardness, softness, springiness, roughness, smoothness, coldness, warmness: they heft, shake, punch, squeeze, push, crush, rub, and try to pull things apart.

—R. Buckminster Fuller
"Children Are Born True Scientists"
1978

Chapter 1

Life Without Touch

In many ways, Chuck Aoki is a typical teenager. At seventeen, he's an avid athlete. His Minnesota Timberwolves basketball team won the national junior championship in 2008. He plays rugby competitively, too. Tennis and baseball are his recreational sports. He likes the history and English courses in his Minneapolis high school, and he is confident that he did well on his SATs, although he hasn't received his scores yet. He likes Taco Bell food. He finds girls confusing. "Women dance around the bushes. Men get straight to the point," he says.[1] He and his friends like to ride the bus to downtown Minneapolis to go to a movie, eat at a restaurant, or hang out at the mall. He's lined up a summer job as a human resources assistant for the Minnesota Twins baseball team, a perfect situation for him since he loves sports. Chuck hopes to go to college on a basketball scholarship. He will major in psychology or sports management.

But for all that, Chuck Aoki isn't a typical teenager. He was born with no sense of touch in his arms, hands, legs, and feet. With a few spots as exceptions, he feels in his limbs neither heat nor cold nor contact nor pressure nor vibration nor pain, although he experiences normal sensation in his head and trunk. His legs were repeatedly damaged in childhood when he walked, ran, and played baseball. His joints have disintegrated and his bones are crumbling. He's full-time in a wheelchair now; his athletic teams are wheelchair basketball and quadriplegic rugby. He's lost eight of his fingers down to the first joint. Text messaging on his cell phone produces bleeding sores on what's left of his thumbs. Buttoning a shirt is impossible. Cutting lettuce for tacos or taking a pizza from the oven poses a serious risk.

Chuck was born with an extremely rare, inherited disorder called hereditary sensory and autonomic neuropathy (HSAN) type 2. This is his story as he shared it with me in a sidewalk café one sunny afternoon. It's also the story of his mother, Jennifer Nelson. She's a librarian for the Minneapolis public library system and the proudest mom I've ever met.

GROWING UP

"You young mothers. You worry too much," the pediatrician told Jennifer when Chuck was a baby. The infant was teething. He'd chewed his fingers to raw and bleeding pulps.

"Is that normal?" Jennifer asked. The doctor wrote a prescription for antibiotics. The baby had an ear infection, he said.

Throughout infancy, Chuck cried intensely and interminably. Only full body contact, his tiny trunk held tight against his mother's skin, could console him. "Is that normal?" Jennifer asked. The doctor wrote more prescriptions for more antibiotics to treat more ear infections.

Chuck walked at nine months, but he walked on his knees. "Is that normal?" Jennifer asked. More antibiotics. Babies get lots of ear infections.

But by the time Chuck celebrated his first birthday, the fact that something was wrong could no longer be blamed on ear infections. The tentative diagnosis of HSAN type 2 came when a neurologist examined a sensory nerve taken from the boy's foot. The fiber was smaller than it should have been, and it lacked the dozens of tiny, hairlike projections usually seen on nerves. The fiber also lacked the myelin sheath that surrounds, protects, and insulates normal nerves. In the absence of myelin, nerve impulses could not travel along Chuck's sensory nerves. The baby's motor nerves were normal, so his brain could direct his limbs to move, but his sensory nerves were nonfunctional. He could feel nothing past his shoulders and his hips.

Chuck was eighteen months old. None of the doctors knew how to treat his condition; they'd never seen anything like it before. The only model came from diabetic neuropathy, which shares some common features with the inherited neuropathies. Diabetic neuropathy compromises blood flow, leading to infections and the "spontaneous amputations" that have robbed Chuck of his fingers, one by one, over the years. Chuck also developed Charcot joints, in which inflammation and loss of healthy bone lead to the overgrowth of abnormal bone tissue, fragmentation, and instability. Why? Because when he walked, he felt the ground only from his hips. He had no feedback to tell him how much force he was exerting with the muscles of his legs and feet, so he stomped too hard. Similarly, when using his hands, his sensation begins at his shoulders. He pushes too hard when he uses a computer, plays video games, or holds a spoon. As a result, his hands constantly bleed and callus. His mother debrides and bandages his wounds daily.

As a small child, he could grasp with his palm but he never learned to make the finely controlled, pinching movements needed to pick up small objects. Because he felt no pain, everyday activities were hazardous. "We pursued a strategy of avoidance," Jennifer says. The hot water temperature was lowered so that Chuck would not burn himself. Jennifer was ever vigilant, keeping the boy away from flames, sharp objects, and hot surfaces.

Jennifer recalls a family vacation to Oregon when Chuck was seven. "We played on the beach all day," Jennifer says, "and that evening Chuck's left knee was swollen to the size of a grapefruit." The child had been seeing a rheumatologist, and Jennifer suspected a flare-up. She called the doctor and asked what to do; the family was planning to take a train home the next day. In the absence of an open wound, infection, or any obvious break, the doctor advised merely watching the situation and keeping the boy off the leg as much as possible on the journey home.

Back in Minneapolis, Chuck went for x-rays. Nothing showed up. He went for an MRI. Nothing appeared wrong except for the persistent inflammation, so Jennifer continued icing and wrapping the joint. Weeks passed with no sign of improvement, so the rheumatologist consulted a specialist radiologist. The radiologist examined the images and found a shadow that previous examiners had missed. An orthopedist then decided to open the leg surgically. He found a fracture of the femur, the large bone of the thigh. "Chuck had been walking around for six weeks on a broken leg," Jennifer says. He was home from school for eight weeks. He had a cast on for twelve weeks. Soon after the cast was removed, he broke his ankle, so he had to be in another cast. He couldn't use crutches because he couldn't coordinate the alternating right- and left-arm actions that crutches require.

When Chuck was nine, doctors discovered that the earlier fracture of the left femur had destroyed most of the bone's growth plate. Subsequent surgery closed the entire growth plate in his right leg and what remained of the growth plate in his left leg to prevent unequal limb growth. "A limb length discrepancy would be dangerous on a daily basis," Jennifer says, "getting in and out of the shower and that kind of thing." After that, he snapped his anterior cruciate ligament (ACL). "We don't know how he did it. We didn't repair it. What would be the point?" asks Jennifer. "His feet are flat, his ankles have sunk into his heels, and both of his feet are deformed," she says. Chuck has been using a wheelchair full-time since he was twelve.

DIAGNOSING HSAN

Throughout Chuck's childhood, his parents remained uncertain about their son's diagnosis. They took the boy to the Mayo Clinic when he was nine in hopes of getting a definitive answer. A precise diagnosis was important, Jennifer explains, because it would help the family plan for Chuck's health and care as he grew. His neuropathy was obvious, but the autonomic part of the HSAN description was not. The autonomic nervous system controls those bodily functions outside conscious control. It regulates heart rate, blood pressure, intestinal action, and glandular function. In all those ways, Chuck appeared normal. Did he really have HSAN? If so, was it really type 2? There are several kinds, each with its own symptoms, course, and prognosis.

To look for the autonomic component of the diagnosis, doctors at the Mayo Clinic did a sweat study on Chuck. "They put him on a little gurney, sprinkled him with powder, and put him into a makeshift, plastic covered oven, and heated the space up," Jennifer explains. "Then they charted where his body turned purple, which is where he sweated. He doesn't sweat from his fingertips and he doesn't sweat below his knees. There was the autonomic part of his neuropathy that we hadn't known about." The HSAN type 2 diagnosis was confirmed.

The broad category of all hereditary sensory neuropathies (HSNs) includes as few as four and as many as six disorders, depending on which classification system is used. All types are genetic, and all involve loss of feeling in the hands and feet. One HSN that is more common and better known that Chuck's disorder is familial dysautonomia (FD, HSAN type 3, or Riley-Day syndrome). Children with FD often experience feeding difficulties and fail to produce tears when they cry. Other symptoms may include lung dysfunctions, cardiac irregularities, poor growth, and scoliosis. Children with FD are most often born to parents of Eastern European Jewish heritage, with an incidence of one in every 3,600 live births.[2] Genetic tests are used to diagnose FD but not the other HSAN types.

Chuck's HSAN type 2 is inherited as an autosomal recessive. That means both Jennifer and her husband, Andy, carry the gene on chromosome 12. Their chance of having a child with the disorder is one in four. (Chuck's younger brother, Henry, is unaffected.) Although the inheritance pattern is understood, the reasons why symptoms vary among individuals are not. Some babies with HSAN type 2 feed poorly, exhibit poor muscle tone, and can't maintain body temperature. Chuck had none of those symptoms, but like most others with the disorder, he suffered numerous fractures of hands, feet, and legs, as well as Charcot joints. Also like most others with the condition, Chuck's muscular strength is normal.

THE BRAIN WITHOUT TOUCH

As for what happens in the brain when touch signals fail to enter it, Jennifer and Chuck's doctors can only speculate. The somatosensory ("body sense") region of the cerebral cortex lies in the parietal lobe at the top of the brain, near the crown of the head. Adjacent to it and forward from it, in the frontal lobe, lies the primary motor cortex, where voluntary actions are triggered. These two long,

Touch the Pain Away

Mothers and infants know what the rest of us may forget—a tender touch takes some of the pain away. A pediatrics team in Boston divided mothers and their newborn infants into two groups. Half of the mothers held their babies in whole-body, skin-to-skin contact while a physician performed the standard heel-stick procedure to draw blood samples from the infants. The other half of the babies were wrapped in receiving blankets and placed in bassinets while the blood was drawn. Babies in contact with their mothers grimaced 65 percent less than the bassinet-held babies did, and their crying time was a whopping 82 percent less. The held babies did not experience the same rise in heart rate as the control babies did. The researchers concluded that skin-to-skin contact reduces both stress and pain for infants.[3]

skinny areas lie sandwiched together, and their nerve fibers are organized in descending rows, like twins lined up side by side. Each nerve and each small area is dedicated to a particular body part—forming side-by-side regions for feeling and movement in the left arm, the right leg, and so on. But the two parallel rows are not identical. In the motor cortex, the amount of "processing power" devoted to a body part varies with the precision of the movements that body part can make. So, for example, the fingers get more space in the motor cortex than the toes do. In the somatosensory cortex, the allocation of space depends on the body part's sensitivity to heat, cold, pressure, vibration, contact, or pain. Thus, the fingertips and lips get a disproportionate share of the somatosensory cortex. The back gets short shrift.

This basic anatomy suggests (but by no means proves) that Chuck's somatosensory cortex may have allocated space differently from the way it is allocated in other children. Receiving no sensory impulses from his limbs, the parts of his somatosensory cortex that would have processed touch information from arms, leg, hands, and feet probably failed to develop. Did the sensory nerves from his head and trunk take over the brain's spare "real estate"? There's no way to know, but it's possible. It's possible, also, that his motor cortex is organized differently, although what happens to the motor cortex if it fails to receive somatosensory feedback is unknown. Still, we can make some guesses in Chuck's case. Athletes who practice their sport and attain a high level of mastery probably devote larger areas of the motor cortex to the body parts they use most. I suspect Chuck's practice of basketball skills and his mastery of his sport have modified his motor cortex and probably his cerebellum as well—for that's where movements that are practiced enough to become automatic are coordinated.

LIVING TO THE FULLEST

Although the brain can reorganize itself to some extent to meet the demands of its owner's life, major brain regions maintain their own specialized functions. As one expert put it, a brain region steals from its next-door neighbor, not the whole town. So Chuck's brain regions for hearing, vision, and the other senses probably aren't any different from anyone else's. But what he's learned to do with his senses is another matter. Chuck says that he relies on vision to manage daily living. He dribbles a basketball expertly (an action some doctors think he should be unable to perform), using his peripheral vision to judge where the ball is and how forcibly he is propelling it. When playing video games, he holds the controller in front of his eyes and peers over the top to see the screen. He's learning to drive a hand-controlled car. He says he uses his peripheral vision to determine where his hands are on the steering wheel, brake, and throttle.

Chuck's vision, strength, and motivation have helped him achieve in sports. He recently started playing wheelchair rugby on an adult team sponsored by the Courage Center in Minneapolis. He's getting good at it. He practiced with the U.S. Paralympics team in 2008. "Maybe I can get named to the USQRA [United States Quad Rugby Association] team. . . . That would be a step toward a world championship team or the Paralympics team," he says.

As for touch, "It's not all it's cracked up to be," Chuck jokes, then adds more seriously, "I've never had it, so I've adapted, and I don't really get what I'm missing. . . . It's not like it has negatively affected me in any really serious way."

"He thrives in his life," Jennifer says. "It is fun to watch him."

Chapter 2

In from the Cold

Argentinean Andrea Salas isn't just pretty. She's flat-out gorgeous. Although no relation to the fashion model of the same name, she's even more beautiful, in my opinion. The glow of a summer sunset streaks her cascade of auburn hair. Her caramel colored skin is flawless. Usually, when she speaks, her doe eyes dance with merriment.

But not today.

Today she is dredging up memories of her brush with death in Antarctica. The deep waters of the world's most extreme environment threatened to take her life. If the sea didn't, the cold surely would have.

ALL ABOARD FOR THE ANTARCTIC

November 23, 2007. Andrea's ship, the *Explorer*, had been at sea for thirteen of its scheduled eighteen days of sightseeing along the margins of Antarctica. Andrea, age thirty-eight, was a veteran Antarctic guide, having served on two previous voyages. This time, as one of sixty-four crew members, she was working as assistant to the leader of the expedition team. She organized lectures, wet landings, and recreational activities for the *Explorer*'s ninety-one passengers, a sixteen-nation aggregate of intrepid tourists who had opened their wallets wide for this, the adventure trip of a lifetime.

The ship had departed from Ushuaia, Argentina, a town that fights a perpetual feud with Chile's Punta Arenas over which is the southernmost city in the world. The itinerary called for a stop in the Falkland Islands—the Malvinas to the Argentineans—and a visit to an abandoned whaling station in South Georgia.

From there, Andrea and her shipmates traveled to Elephant Island, famous as the last refuge of the men of Shackleton's *Endurance* Expedition of 1914–1916. Shackleton failed to cross the Antarctic continent on foot as he had planned, and he was forced to leave his men on Elephant Island while he piloted a single lifeboat across eight hundred miles of open ocean to find help on South Georgia. The landscapes of the islands were breathtaking, Andrea recalls. Wildlife is scarce in this, the most inhospitable environment on earth, but the tourists marveled at the albatrosses of Steeple Jason in the Falklands. They oohed and aahed over the king penguin rookeries of South Georgia.

November is early for an Antarctic trip. The sun doesn't rise in Antarctica until September, and summer doesn't arrive until January when the weather warms to a balmy –15° to –35°C (–5° to –31°F). The weather was exceptionally cold on this trip. Andrea kept her charges wrapped in layers of thermal underwear, polar fleece, and waterproof jackets. Boots swaddled feet, while waders cocooned passengers toes-to-chest for wet landings. The season and the cold brought with them some disappointments for the passengers and crew of the *Explorer.* A planned side trip to South Orkney Island had to be canceled because the seas were beginning to crowd with icebergs.

It might have been a portent, but the *Explorer* steamed on.

On this Thursday night, Andrea left her cabin on the crew deck, below sea level. There, while resting in bed, she could hear ice hitting the hull. She wasn't worried. The *Explorer* was built to handle icy seas. She headed to the bar to enjoy a drink with fellow crew members and some passengers. The bar was warm; the drinkers, convivial.

ABANDON SHIP!

Around midnight two passengers rushed into the bar, shouting, "There's water! There's water!" Everyone ran from the bar, only to collide with the captain who was ascending the stairs. "The captain told us water was coming in through a hole," Andrea says. She raced to her cabin, as did the others, pulling on warm clothes and collecting her life jacket, while the captain announced from the bridge that emergency evacuation was commencing. *"This is not a drill!"*

"We got to the muster stations almost immediately," Andrea reports. Huddled there in the dark and biting cold, passengers listened to reassurances from the captain over the public address system. Attempts were underway to repair the hull, but just in case, emergency rescue stations in Chile and Argentina had been radioed, as had other ships in the area. Andrea joined her fellow crew members in distributing hot drinks and hope while awaiting the order to abandon ship. It wasn't long before that order came. Repair attempts had failed, and seawater was pouring in through a breach in the hull. Its cause? Probably impact with an iceberg.

Now came the greatest challenge: Andrea and 153 others would have to brave the turbulent Antarctic seas and the bitter cold in nothing more than small lifeboats and Zodiacs, the inflatable rubber boats usually employed for tendering to shore from the ship and sightseeing along rocky coasts and beaches. Andrea hustled passengers into the emergency craft, making sure that each boat had a crew member to handle the motor and steering.

She and two other crew members lowered one of the Zodiacs into the water and scrambled aboard. Zodiacs are not lifeboats, but in Andrea's opinion, they represented a better choice for escape than did *Explorer*'s aging orange lifeboats, which were heavy, clumsy, and poorly powered, although provisioned with food and water. The Zodiacs have powerful motors; they are light and are highly maneuverable. Still, they are open, small, and cramped when holding more than four or five people.

Andrea's Zodiac maneuvered next to one of the crowded, heavy lifeboats, transferring seven passengers into the rubber inflatable. When all the boats were in the water, the crew lashed the lifeboats and Zodiacs together with ropes, using the Zodiacs' engines to keep the lifeboats from ramming into icebergs. The passengers and crew of the *Explorer* had radios that allowed communication among themselves but not with the outside world. They were alone and adrift in the Antarctic seas.

The ten people on Andrea's Zodiac crouched under blankets, settling in for a long wait. Mostly they all stayed silent, the smell of apprehension heavy in the air. "Those were the longest hours of my life," Andrea says. The wind whizzed past her ears. The cold crept through her seawater-soaked clothing to ice her bones. In conditions like these, she discovered, waterproof jackets aren't waterproof.

"My feet and hands were very cold," she says. "I never feared death, but I did fear that I might lose a hand, a finger, a foot. I don't know how cold it has to be to lose a part of your body, but I was afraid it would happen because I knew I would not get any warmer." She and her compatriots were stacked in the Zodiac like cord wood. They had no way to move, no means of generating or sharing heat. "I couldn't feel my hands inside my gloves," Andrea says. "My feet were blocks of ice."

COLD-SENSING

While Andrea waited for rescue, she felt overwhelmed by a paralyzing sense of cold in every part of her body. What gave her that feeling? What was going on in her nervous system to create that perception? The answer lies in tiny receptors on the membranes of neuronal projections that lie close to the skin's surface.

It's long been known that touch neurons specialize. Some respond only to pressure; some, only to light touch. Several subsets of neurons fire an impulse when the ambient temperature rises. Others detect only a decline—or cooling.

(Any temperature lower than that of the human body, 37°C or 98.6°F, is "cold" in the physiological sense.) One type of neuron fires at relatively warm temperatures, 15° to 4°C (59° to 39°F); yet another type activates at temperatures below 0°C (32°F).[1] Still another kind generates an impulse only when cold is so intense that it's painful.

For a long time, no one understood what was going on in temperature-detecting neurons, but that changed in 2002 when several research teams delivered up two prizes: receptors on nerve cell membranes that respond to cool-feeling chemicals such as menthol; and cold-activated ion channels in cell membranes. The receptors work just as taste and smell receptors do; after locking onto a molecule of a cool-feeling chemical, they provoke some change in the cell that, in turn, triggers a nerve impulse. The channels do precisely what their name suggests. They either open in response to cold, allowing ions (charged atoms or molecules) to flow into the cell,[2] or they close to prevent them from leaking out—maintaining the cell's positive charge.[3]

The influx of positively charged sodium or calcium ions (or the maintenance of a positive charge) causes electrical potentials to change across the outer membrane of the neuron. That happens because a neuron at rest has a slightly negative charge inside it, compared with a positive charge outside. When cold causes an ion channel to open, sodium ions flow into the cell, thus changing the charge along its membrane. For a tiny fraction of a second, the inside of the cell becomes positive; the outside becomes negative. This reversal in one tiny area creates a current that affects the membrane farther along. Channels there open, and sodium flows into another section of the nerve fiber, then another, then another. That's how the impulse travels—eventually to the brain.

Receptors and ion channels can work alone or in concert, in the same cell or in different cells. One receptor channel called TRPM8 has been studied extensively. Also called CMR1 (cold and menthol receptor 1), it's a major player in detecting gentle cooling. Researchers at the University of California at San Francisco found its genetic base when they isolated cold-sensitive, menthol-sensitive neurons from rats. They took the *TRPM8* gene from those cells and placed it into cultured neurons that were neither cold- nor menthol-sensitive. The cells changed their function and became cold- and menthol-detecting neurons. Why? Because the gene directed the construction of TRPM8 receptors on the cell membrane.[4]

Another team of researchers that same year discovered how cells that have TRPM8 receptors respond to temperatures in the 23°C to 10°C (73° to 50°F) range. The calcium concentration increases inside the cell. This suggests that TRPM8 is both a receptor and an ion channel.[5] The scientists don't know how cold affects the channel, but they describe two possible mechanisms. Cold might change the shape of the channel directly, or it might trigger the release of a chemical (a "second messenger") that, in turn, prompts TRPM8 to open.[6]

Does Cold Make Your Family Sick?

Familial cold autoinflammatory syndrome (FCAS) is so rare, only six hundred people in the world have it. Its name is something of a misnomer. It doesn't take severe cold to trigger an attack. Some FCAS patients say that merely walking by an open freezer in a grocery store is enough to bring one on. Ordinary room temperatures are too cold for many FCAS sufferers.[7]

People with FCAS experience a constellation of symptoms that may include fever, chills, hives, rash, nausea, eye irritations, headache, muscle aches, and pain, stiffness, and swelling in the joints. "It's like you're freezing from the inside out," says FCAS patient Rachel Doherty. "Your whole body contracts until you warm up, which can take hours or days."[8]

The disease is inherited; the dominant gene for the condition passes from parent to child. If either the father or the mother has it, a child's chance of getting it is (usually) one-half. The cause of FCAS is often a mutation in the gene *CIAS1* on chromosome 1—although other genes may be involved. The disorder begins in infancy and may worsen over time. Despite the disease's debilitating symptoms, most people with FCAS lead normal lives.

Until 2008, the only treatment for FCAS was warming, along with nonsteroidal antiinflammatory drugs (such as ibuprofen) or corticosteroids. In that year, clinical trials showed success with drugs that trap or block the immune-system chemical interleukin-1, which plays a role in the inflammatory reaction.[9]

Another possible mode of action has been explored by researchers in Spain. They studied "voltage-gated" potassium channels in cell membranes. Neurons *in*sensitive to cold have them, but cold-sensitive neurons do not. The channels function as "excitability brakes" during cooling in insensitive cells. They stop an impulse from beginning. Cold-sensitive neurons lack such voltage gates, so temperature changes can initiate an impulse in them. This relationship was established when the Spanish team blocked the action of the voltage-gated potassium channels in cold-insensitive neurons. About 60 percent of the neurons began firing impulses in the cold.[10]

The human body can differentiate a temperature change as small as a single degree Celsius, but only a few heat- or cold-sensitive ion channels like TRPM8 are known. How can relatively few cellular structures account for such a finely tuned capacity? The answer may lie in the ability of the channels to shape-shift. In cell cultures, channels that respond to *increases* in temperature are known to rearrange themselves by combining with other channels within their same structural "family" (not TRPMs but another category called TRPVs). The resulting complexes demonstrate characteristics intermediate between those of the "parent" forms. Research teams can measure the intermediate electrical charges and ion-gating properties of these "hybrid" channels. This ability to shuffle and recombine means that the number and variety of channel types are large. Although channels that respond to *decreases* in temperature have not been researched, it's likely that they, too, combine with others to form intermediate types. Whether such channel complexes act as functional temperature sensors in living organisms remains unknown.[11]

RESCUE!

Andrea never gave a thought to her TRP receptors. She had survival on her mind.

As the wee hours of the morning gave way to a (blessedly) clear dawn around 3 A.M. (nights deliver less than three hours of darkness at this latitude in November), Andrea and nine others watched from the safety of their Zodiac as the *Explorer* began to list starboard. The *Explorer*'s engines had failed, perhaps swamped by water. The boat careened drunkenly with the swirling currents, powerless and empty after the captain was the last to abandon his post. *Explorer* drifted toward seas increasingly jammed with icebergs, driven by a rising wind. Before it sank at the end of the day, the *Explorer* would be completely surrounded by pack ice.

Andrea remembers the noise. "The sound of a ship when it is breaking and about to sink is just like in the movies," she says, her dark-chocolate eyes wide with the memory. "I could hear pipes breaking." The eerie creak of twisting metal assaulted her ears unceasingly as what passes for an Antarctic dawn brightened into morning.

And yet they waited for hours more, hoping against hope for a rescue they could not be sure would come.

Then the Norwegian cruise ship, the *Nornodge*, topped the horizon. *Rescue!* The Zodiacs raced toward the *Nornodge*, with the slower lifeboats trailing behind. "We didn't mind getting wet anymore," Andrea says, her eyes brightening now that she can report a happy ending. Miraculously, not a single person suffered injury—not even frostbite. The *Nornodge* doctor offered treatment to the rescued sailors—all 154 of them. Only two took advantage of the offer.

Andrea recalls how her sense of lethal cold disappeared when the rescue ship arrived. "When I stood to look at the *Nornodge*, all the heat returned to my body," she says. "The hope I gained from seeing it sent a wave of warmth through me. It was the excitement of feeling safe. Knowing that it was over, I didn't feel cold anymore."

Chapter 3

On Which Side Is Your Bread Buttered?

Do this: Go to the kitchen and take out the bread and butter. Now butter a slice of bread, but, for the first time in your life, don't do it automatically. Think about what you are doing. Since nerves from the left and ride sides of the body cross when they enter your brain, your left brain controls your right hand and vice versa. If you are like most people, you'll find yourself using your preferred or dominant hand (probably the one you write with) to butter the bread, while your other hand holds the slice. Now notice something else. Again, if you are like most people, you'll look at the butter as you spread it. You'll pay no attention to the bread and the nondominant hand that's stabilizing it. Thus, if you are right-handed, your left brain is controlling the knife and monitoring the butter-spreading visually. Your right brain stabilizes the bread with your left hand, monitoring the force and position of the slice with your sense of touch. (It's the opposite if you're left-handed.)

So, you may be thinking, *what's the big deal?*

It turns out that how you butter a slice of bread reveals a great deal about how the brain processes sensory information. Put away the bread and butter and fly off with me to the University of Michigan in Ann Arbor to learn more.

PROBING PROPRIOCEPTION

Where else would a university house a division of kinesiology but in a 1930s Tudor revival-style building straight out of a gothic novel? Standing incongruously within spitting distance of the University of Michigan's ultramodern hospital complex, the red brick Observatory Lodge is heavily ornamented with stucco veneers, angled half-timbers, hexagonal bay windows, and more gables than seven.

Behind one of those gables, in a fourth-floor aerie that serves as laboratory, office, and student research center, I find motion scientist Susan Brown. She's a warm, friendly woman with red hair and dancing eyes, stylishly and colorfully dressed in contrast to the spare, starkly functional room where she works. But there's no need for decoration here. There's work to be done. Several students busy themselves at computer stations around the room, eyeing me curiously.

Brown has prepared for my visit. She's set up a webcam at one of the computers so we can converse with her former graduate student, Dann Goble, now a postdoctoral fellow at the Research Center for Movement Control and Neuroplasticity at Katholieke Universiteit Leuven in Heverlee, Belgium. It's evening in Belgium, and youthful, soft-spoken Dann relaxes with a Stella Artois as he describes how his research began when he first started working with Brown in his student days.

"We were running some simple experiments," Goble says, "looking at how well people can match position when they don't have vision." For those experiments, Goble blindfolded his subjects and placed their arms on supports, elbows out and hands on levers. One arm then moved to a certain spot and back again. Sometimes Goble moved his subjects' arms for them. Sometimes the volunteers initiated their own movements. In either case, the task was simple: match the movement, either with the same arm or the opposite arm, hitting the same spot as closely as possible using not vision, but the proprioceptive sense—the sense of where the body is in space.

"For many years," Goble explains, "nearly all the experiments that were carried out in movement science looked at how the dominant arm works. But we decided for some reason to test both arms, and we obtained this really strange result: The nondominant arm was working better than the dominant arm." Goble doubted his own findings. Maybe he wasn't performing the test correctly. He checked and double-checked, but nothing seemed to be wrong with the procedure, so Goble went back to the research literature to see if anyone before him had reported such a finding. Sure enough, in a long-overlooked paper from nearly thirty years ago, he found a similar result, not for the arm but for the thumb. Research subjects had once matched thumb positions better with their nonpreferred hand than with their preferred,[1] but no one in the research community had paid much attention to such an oddity. *The dominant arm and hand are always better at everything, aren't they?* Goble's findings flew in the face of "what everybody knows." "That's when we thought maybe we were really on to something," Goble says.

Together Goble and Brown set up a series of experiments to explore "laterality," or the distinction between the operation of right and left arms. If what they were beginning to suspect were true, their research could take them far beyond the

Is Proprioception the Same as Touch?

Proprioception is the sense of the body's position in space—and of its movement, speed, and force. Brown and Goble say proprioception and touch are more or less the same thing, although proprioception may be thought of as a particular type of touch. The receptors that initiate impulses of pressure, vibration, and temperature lie in the skin. The receptors more important for proprioception lie deeper—in the joints, tendons, and muscles. The muscle spindle, found in the body of a muscle, is one type of proprioceptor; it signals changes in muscle length. The Golgi tendon organ, which lies in the tendons that attach muscle to bone, provides information about changes in muscle tension. These receptors send their impulses to the somatosensory cortex of the brain. It lies in the parietal lobe (the top of the head). Proprioceptive information also travels to the cerebellum, where automatic and habitual motions are controlled.

oddities of matching thumbs and arms, even beyond the daily rigors of buttering bread. Could it be, they wondered, that the two arms differ in their preferred sensory inputs, with the dominant hand attuned to vision and the nondominant acting as the primary source of proprioceptive information?

The researchers began testing their "sensory modality hypothesis of handedness" with experiments that allowed proprioceptive information only. Blindfolded volunteers worked under three separate conditions, each a little more complex than its predecessor in what the brain was required to do:

Task 1: same arm match. One arm (either left or right) was moved to a target position between 20 and 60 degrees of elbow extension and held for two seconds before being returned to the starting position. The task was to move that same arm back to that same position, using only proprioceptive memory on a single side of the brain (because it was the same arm).

Task 2: opposite arm match with reference. One arm was moved in the same way as in task 1, but it was left in the target position, so it could serve as a stable marker. The charge was to move the opposite arm to the same position. This task required no proprioceptive memory, but it was more difficult, because it required communication between one side of the brain and the other—matching one arm to the other.

Task 3: opposite arm match without reference. As in task 1, one arm was moved to a target position, held for two seconds, then returned to the starting position. The task was to repeat the movement with the opposite arm. This was the most difficult task of all because it required proprioceptive memory as well as communication between the two sides of the brain.

On all three tasks, the nondominant arm (the left for most people) was better at matching positions. Furthermore, the left arm/right brain superiority grew with the difficulty of the task. Task 3, the opposite arm match without reference that required both memory and cross-brain communication, showed the biggest difference of all. The nondominant arm's performance was even better than on the two simpler tasks, which required either memory or the brain's cross-talk, but not both.[2]

Brown and Goble conducted another series of experiment to compare performance when visual cues were provided. As expected, the dominant arm excelled,[3] but when only proprioception was allowed, the nondominant arm won every time. Over several years, Brown and Goble tried other approaches. They tested children. Same result. They tested the elderly. Same result. They conducted experiments in which volunteers matched not position, but force. Still, the results were always the same. In a right-handed person, the left arm/right brain partnership is a champion proprioceptive team.

LATERALITY

Ask one hundred people which is their preferred hand, and nine out of ten will say their right. Apparently, the preference for right-hand, right-arm activity has been around as long as humans have been throwing spears and painting pictures on cave walls. The right-handed use of stone tools appears in images that date back 35,000 years. Since the left brain controls the right hand, it's tempting to think that the left hemisphere is more or less in charge of most skilled movements, since we think we don't do much with our nondominant (left) arm, and some researchers find evidence that that's true. One study showed that the right motor cortex actively controls movements of the left hand, but that the left motor cortex is active in movements of both hands.[4] No one can explain that finding, but it's possible that the left hemisphere sends messages to the left hand (by way of the right hemisphere) via the corpus callosum, the stalk of brain fibers that connects the brain's two halves.

While it's true that the preferred hand is better at many tasks, Brown and Goble have shown that it's probably not the major source of proprioceptive feedback to the brain—at least not when the two arms must perform in different, yet complementary, ways. They aren't the only scientists to reach that conclusion. Researchers had found that patients with right-hemisphere brain injuries have more trouble reproducing movements than do those with left-hemisphere lesions.[5] Some brain-imaging studies have compared brain activity while subjects reached for remembered targets. In some cases, the targets were visual; in others, they were proprioceptive. During the proprioceptive tasks, the right hemisphere's

Pinocchio's Nose

Vibration disturbs the proprioceptive sense, and the illusion of Pinocchio's nose proves it. Here's how it goes: Close your eyes and place the tip of your finger on your nose. Your brain immediately draws conclusions about the length of your nose. But if someone were to vibrate your elbow joint, you'd feel that your nose had suddenly grown longer. Why? Because vibration stimulates muscle spindles, misleading the brain into concluding that the vibrated muscles have stretched and the arm has moved, although the limb has actually remained immobile. The brain draws the only logical conclusion it can: If your arm muscle has stretched and if your finger is still on your nose, your nose must have grown longer.

"Vibration is a great tool for fooling the nervous system," Susan Brown says, and researchers have used that fact to uncover further evidence for right-brain dominance of the proprioceptive sense. An international team of researchers from Sweden, Japan, German, and the United Kingdom vibrated tendons in the arms of right-handed, blindfolded volunteers. The vibration made the subjects feel that their fingers were curling in toward their palms, although, in reality, their hands remained fixed. The scientists used functional magnetic resonance imaging (fMRI) to pinpoint the brain's response to the perceived, but unreal, motion. They found that the proprioceptive regions of the brain's right hemisphere grew much more active than those of the left, no matter which hand was thought to move. "Our results provide evidence for a right hemisphere dominance for perception of limb movement," the researchers concluded.[6]

somatosensory area showed greater activity than did the left, even when all the sensory information came from the preferred right arm.[7]

ASYMMETRIES IN THE REAL WORLD

The specialized, yet cooperative, partnership between the two arms has obvious applications in daily life. In right-handed people, the left hemisphere's affinity for vision promotes fast, accurate, visually guided action of the preferred right arm, as when reaching for a slice of bread. The right hemisphere's specialization in proprioceptive feedback helps the nondominant, left arm maintain the position of an object, such as holding a slice of bread, while the preferred right arm acts on it—in this example, spreading the butter.

Brown's and Goble's research helps us better understand how our brains, arms, and hands work, but it has practical applications, too, in rehabilitative treatments for those with movement and proprioceptive disabilities. Goble has worked with children who have cerebral palsy. He's tested them on matching tasks and found what he expected. Children with right hemisphere damage are usually worse at proprioceptive matching tasks than those with damage on the left side. Brown has been testing some home-based training protocols with adults who have cerebral palsy. The training requires subjects to perform small tasks, such as picking up objects with chopsticks or turning over cards. She tells of a man with cerebral palsy who, now in his early thirties, has always reached across his body with his left hand to put the key in the ignition of his car. After only eight weeks of home training, he can now use his right.

"We have a mix of both right-side-affected and left-side-affected [people] in that study," Brown explains, "but they are all getting the same intervention. . . . We don't have enough numbers, but we wonder if we are getting more of an effect for those that might have damage to one side of the brain because it's more responsive to the types of training we are using." Brown is also starting a collaboration with the University of Michigan Stroke Clinic to develop training regimens specifically designed for right- and left-side impairments. "We have enough basic science data, but now we need to translate that into a clinical environment," she says. A great deal has changed in treating patients since researchers learned how the brain rewires and restructures itself after an illness or injury. "It's a big deal now in clinical rehabilitation to pay attention to sensory feedback and not just concentrate on getting the brain to contract muscles to produce movement," Brown says. "Brain plasticity has caused a real paradigm shift in rehabilitation."

Chapter 4

Pain and the Placebo Effect

It's tempting to think that our senses provide us with an accurate representation of the external and internal environment. We believe that our eyes project into our brains a perfect three-dimensional video recording of the objects and events around us. It's easy to surmise that our ears detect and transmit all incoming sound waves with sound-studio clarity, and that our mouths, noses, and fingertips provide us with all the evidence we need to draw indisputable conclusions about tastes, odors, and textures.

We are kidding ourselves. Our senses respond to only a fraction of what's out there, and they respond in ways so malleable and idiosyncratic as to call into question the whole concept of truth. Consider the legendary unreliability of eyewitness accounts. In a recent study, Canadian researchers showed a video clip of a movie action sequence to eighty-eight undergraduates who then discussed what they had seen either with one or with several of their peers. In the debriefing sessions, a researcher, posing as a peer, deliberately introduced misinformation into the discussion. Later, on true-false tests of facts about the movie scene, participants in the one-on-one condition accepted the impostor's false suggestions 68 percent of the time; participants in the larger groups were taken in 49 percent of the time.[1]

Perhaps our eyes, ears, and memories can be fooled, you may think, but internal pain is different. We know where we hurt, how much we hurt, and how we feel when the pain increases, diminishes, or disappears. Pain is a very personal and intimate sense, and there is no sidestepping it, denying it, or talking your way out of it. You know when you hurt!

Or do you?

What Is Pain?

An estimated 50 million Americans experience persistent pain, often from backache, headache, arthritis, or cancer.[2] Like everything else in the brain, pain is a matter of physics and chemistry. The response starts with stimulation of nerve fibers called nociceptors. They lie in the skin, muscle, and other body tissues. Their impulses carry messages to the thalamus and cerebral cortex of the brain where pain signals are processed. Pain signals also pass to other brain regions such as the rostral ventromedial medulla (RVM), where the "volume" of the pain message can be turned up or turned down through signals sent from the RVM to the spinal cord.[3]

Unlike the nerve endings that respond to light touch or pressure, nociceptors ordinarily need a strong stimulus to begin firing. Inflamed or injured tissues, however, release chemicals that make nociceptors more sensitive so that a weaker stimulus causes them to fire. Unlike many other kinds of receptors, which quit triggering a signal when stimulated for too long, nociceptors become increasingly sensitive with continuing or repeated stimulation.

All this action is under genetic control, and so is the experience of pain. Researchers at the University of Michigan and the National Institutes of Health have found that a variation in a single gene helps to explain why some people can tolerate more pain than others can. The gene has two forms called *met* and *val*. Both code for the enzyme COMT, which breaks down the neurotransmitters dopamine and noradrenaline. (Neurotransmitters are the chemicals that ferry an impulse from one neuron to the next.) COMT removes neurotransmitter molecules from the space between neurons when an impulse is finished.

However, because the genes *met* and *val* are different, they make the enzyme in slightly different forms. People who have two copies of the *met* form have a much greater response to pain than those who have two copies of the *val* form. Those who have one copy of each form have pain tolerance somewhere in between. The *val* form, it seems, makes an enzyme that does its job efficiently and rapidly, reducing the experience of pain for its lucky owner.[4]

THE PLACEBO EFFECT

Let's do a thought experiment. Round up three hundred harried commuters with headaches—not hard to do on the New York subway any workday rush hour. Of course, they are shouting and whining strident protests, which only worsen their headaches, which is precisely what you want. You reassure them that you'll get their names mentioned in the *New York Times* in recognition of their public service (you can't afford to pay them), and that settles them down long enough for you to herd them into three soundproof rooms, one hundred headaches per room.

Now the fun begins. You do nothing with the first one hundred. They get to glare at one another Big-Apple-style and ruminate on their throbbing temples. You make an eloquent speech to the second group, informing them that they are the lucky recipients of a newly developed and powerful painkilling miracle drug. (It's actually aspirin with codeine, a proven pain reliever.) Then you leave them, too, alone with each other and their pain, contemplating their lawsuits against you. You make the same speech to the third one hundred, but you are lying to them. They think you are giving them a pain-relieving drug. In truth, they get a sugar pill.

After a half hour, you ask your three hundred captives to report on their headaches. In the "do nothing" group, twenty say their headaches are gone. Eighty are still suffering. In the second group, ninety report the complete disappearance of pain; that drug is certainly a miracle potion, the people say, and they wonder where they can purchase it. In the third group, the ones you deceived, forty-five still have headaches, but fifty-five do not. That pill did the trick, they say, happily reboarding the subway pain-free. Your experiment was a success and you are off the hook, unless one of your subjects is a liability lawyer.

But forget the legal ramifications for now. Look at what the experiment revealed. A sugar pill has no physiological action that will cure a headache, but thirty-five of your headache-free subjects in the third group provide evidence to the contrary. (Why thirty-five and not fifty-five? Because the results from the first group, the "do nothing" group, show headache pain will cease in 20 percent of your subjects after a half-hour regardless.) Thus, for 35 percent of the subjects in our thought experiment, the sugar pill was just as much a miracle drug as the painkiller the members of the "real drug" group received. This "cure" in the absence of any truly therapeutic agent is the placebo effect, and it's more than a curiosity. It's a direct result of brain action. But how?

Before we answer that question, we need to define precisely what the placebo effect is. It is not spontaneous remission. That's what the twenty people in the first group (and presumably twenty more in each of the other two groups as well) experienced. Some of us, no matter what the disease, get better for unknown reasons. The disease process simply reverses itself without any intervention. Whether remission is mere chance or the result of some self-healing process remains anybody's guess.

Neither is the placebo effect deception or self-delusion. The people whose headaches disappear after ingestion of the sugar pill are not lying, cheating, simple-minded, or insane. Their pain disappears—and not because they consciously wish it to. In study after study, where both subjects and experimenters are "blind" to the experimental conditions—that is, no one, including the researchers, knows who is getting the placebo—measurable, clinically replicable improvements in disease conditions occur in a sizable fraction of all cases.

Furthermore, the placebo effect is no small or insignificant statistical aberration. Researchers at Brigham and Women's Hospital in Boston performed some mathematical magic to combine the results of nineteen separate studies (conducted between 1970 and 2006) on patients with irritable bowel syndrome. Across all studies, the placebo response averaged nearly 43 percent, ranging from a low of 15 percent to a high of 72 percent. The longer the period of treatment and the larger the number of physician visits, the greater the placebo effect.[5]

Finally, the placebo effect is not restricted to subjective self-reports of pain, mood, or attitude. Physical changes are real. Working under a grant from the National Institutes of Health, a multistate team of researchers tested fifty-five asthma patients who had recurrent mild to moderate symptoms. Patients were randomly assigned either to treatment (salmeterol) or to placebo groups, and all participants were blind to the test; neither the patients nor their doctors knew who was getting the placebo and who was getting the drug. The researchers assessed not the patients' self-reports of symptoms, but the degree of constriction in their bronchial tubes. A one-dose trial showed the placebo effect in more than half the patients. A stringent two-dose trial narrowed the field to 18 percent of subjects who were unequivocal placebo responders. Five of the subjects (9 percent) were completed "cured" by the placebo treatment.[6]

The Placebo Effect in the Brain

The placebo effect is not deception, fluke, experimenter bias, or statistical anomaly. It is, instead, a product of expectation. The human brain anticipates outcomes, and anticipation produces those outcomes. The placebo effect is self-fulfilling prophecy, and it follows the patterns you'd predict if the brain were, indeed, producing its own desired outcomes. For example, researchers have found the following:[7]

- Placebos follow the same dose-response curve as real medicines. Two pills give more relief than one, and a larger capsule is better than a smaller one.
- Placebo injections do more than placebo pills.
- Substances that actually treat one condition but are used as a placebo for another have a greater placebo effect that sugar pills.
- The greater the pain, the greater the placebo effect. It's as if the more relief we desire, the more we attain.
- You don't have to be sick for a placebo to work. Placebo stimulants, placebo tranquilizers, and even placebo alcohol produce predictable effects in healthy subjects.

- Placebo effects can be localized. In one experiment, Italian researchers injected capsaicin into the hands and feet of brave volunteers. Capsaicin is a derivative of hot peppers; it produces a burning sensation. The wily scientists then applied a cream to only one hand or foot and told their subjects that the cream was a powerful local anesthetic. In truth, the cream was an inactive placebo. Nevertheless, the subjects reported less or no burning on the treated hand or foot, but the burning sensation was strong elsewhere.[8]

As in all brain actions, the placebo effect is the product of chemical changes. The first clues to the chemical basis of the placebo effect came in 1978, when research on pain in the dentist's office showed that the drug naloxone can block placebo-induced pain relief.[9] Naloxone is an opioid antagonist; it locks onto the brain's natural painkillers, the endorphins, and stops them from doing their job. Numerous studies have supported the conclusion that endorphins in the brain produce the placebo effect. In patients with chronic pain, for example, placebo responders were found to have higher concentrations of endorphins in their spinal fluid than placebo nonresponders.[10] One clue that expectation triggers the release of endorphins came from other studies of naloxone conducted the 1990s. Italian researchers reported that naloxone blocked the placebo effect produced by strong expectation cues; but in the absence of expectation, the drug had no effect.[11]

Some researchers and theorists debate whether expectation or conditioned learning accounts for the placebo effect, but it's not hard to construe conditioning as a special case of expectation. Conditioned learning is the classic Pavlovian response. When dogs were fed at the same time that a bell was rung, it wasn't long before the dogs started to salivate at the sound of the bell, although no food was present. The neutral stimulus (the bell) became associated with the effect-producing stimulus (the food). The same mechanism might explain some forms of the placebo effect. If a pill or injection (neutral stimulus) has been associated with a physiologically active drug (effect-producing stimulus) in the past, then later the pill alone may produce the action without the drug. For example, your body won't make any more growth hormone just because your doctor tells you it will. But if you take the drug sumatriptan, which actually *does* increase growth hormone secretion, and later you take a placebo that you *think* is sumatriptan, guess what? Your body will make more growth hormone when you take the placebo drug, even if your doctor tells you it won't.[12]

That is an example of the placebo effect that involves neither pain nor the brain's production of endorphins, but, if conditioning is considered a form of expectation, it makes the same point. Your brain will direct your body to act as it's expected to act—whether the expectation is consciously perceived or unconsciously learned.

Placebos for Emotional Pain

Expectation is strong medicine for a variety of physical and emotional ills, even our perceptions of horrific images and events. A research team led by Predrag Petrovic of the Karolinska Institutet in Stockholm showed emotionally loaded pictures, such as images of mutilated bodies, to fifteen healthy, adult volunteers. On the first day of the two-day experiment, the researchers gave the volunteers an antianxiety drug to dampen their subjects' emotional response to the pictures. Then the scientists gave an antidote to the drug and told the volunteers that the antidote would restore the full intensity of their emotional reactions.

The next day, the subjects were told they would take the same drugs, but they were deceived. They got nothing more than a placebo salt solution when they viewed the pictures. The placebo reduced the subjects' negative reactions to the unpleasant pictures by an average of 29 percent. Functional MRI scans of the volunteers' brains showed that the placebo reduced activity in the brain's emotion-processing centers. The effect was mathematically predictable; as brain activity diminished, so did the perception of negative emotions in response to the pictures. The scans also showed increased activity in the same brain regions known to respond with the placebo effect in reducing the perception of pain. The effect was greatest in those subjects who reported the largest effect from the real antianxiety drug administered on the first day.[13]

What's the point of administering a real drug the first day and a placebo the second? This experimental design treats the placebo effect as a case of conditioned learning. If you've experienced a positive benefit from a treatment in the past, your expectation of its benefit in the future is stronger, and so is the placebo effect.

Making the Most of the Placebo Effect

At one time, researchers viewed the placebo effect as an impediment—a statistical annoyance that got in the way of objectively evaluating the efficacy of potentially legitimate therapies. That view has changed. The placebo effect is today seen as an important part of the healing process. It's been studied as a treatment for Parkinson's disease, depression, chronic pain, and more. For large numbers of patients—the placebo responders—belief in the therapy will create or enhance its effectiveness.

Life in Chronic Pain

People who suffer constant pain carry with them a load of other miseries: depression, anxiety, sleep disturbances—even difficulty making decisions. Dante Chialvo, a researcher at Northwestern University, has found a reason why.

Chialvo's research team used functional magnetic resonance imaging (fMRI) to scan the brains of thirty adult volunteers. Half the volunteers had chronic low-back pain. Half were pain-free. The subjects used a finger-spanning device to rate continuously the height of a bar moving across a computer screen.

In the pain-free group, activation of some parts of the cortex while performing the task corresponded with deactivation of other parts—a state Chialvo calls a "resting state network." The chronic-pain subjects performed as well on the task as their pain-free counterparts, but their brains operated differently. One part of the network did not quiet down as it did in the pain-free subjects.[14]

A healthy, pain-free brain, Chialvo asserts, maintains a state of "cooperative equilibrium." When some regions are active, others quiet down. But in people who experience chronic pain, many brain regions fail to deactivate. A region of the frontal cortex associated with emotion "never shuts up," Chialvo says.

This constant firing of neurons could cause permanent damage. "This continuous dysfunction in the equilibrium of the brain can change the wiring forever," Chialvo says. "It could be that pain produces depression and the other reported abnormalities because it disturbs the balance of the brain as a whole."[15]

Placebo treatments for Parkinson's disease offer a good example. Parkinson's is a decline in the brain's ability to control movement. The disease causes a slow, shuffling walk, stiff shoulders, and trembling. Parkinson's symptoms result from the death of neurons in several regions of the brain. The affected neurons normally control motion by releasing the neurotransmitter dopamine. Cell death means that people who have Parkinson's have too little dopamine in their brains. One recent study of placebo treatments used positron emission tomography (PET) to measure the brain's manufacture and release of dopamine. Predictably, the brain produced more dopamine when patients were given a placebo and told to expect an improvement in their motor performance.[16]

In some respects, the placebo effect offers the best of all possible alternatives: therapeutic effects without the risk of negative side effects. It does, however, have its downside. Just as expectation can produce positive outcomes, it can also produce negative ones. That's the "nocebo" effect, named from the Latin for "I shall harm." A harmless—but dreaded or feared—treatment induces a new illness or exacerbates an existing one. Practitioners of black magic and voodoo curses have known that for centuries, but the nocebo effect is difficult to research for obvious reasons. It's unethical to threaten people with illness, but a few brave investigators tried it before institutional review committees began to place stringent constraints on research involving human and animal subjects. For example, in 1968, scientists asked forty nonasthmatic persons and forty asthmatics to inhale water vapor purported to contain potent irritants and allergens. The nonasthmatics had no reaction to the vapor, but nearly half of the asthmatics did. They experienced a measurable narrowing of their airways. Twelve of them responded with asthma attacks that were successfully treated with the same saline solution, which they were led to believe contained an effective antiasthma drug.[17]

Doctors could do more with the placebo effect if they could understand why it works for some people but not others. Part of the answer is—once again—expectation. Jon-Kar Zubieta of the University of Michigan at Ann Arbor used PET to study the brains of men who had jaw pain. Zubieta gave them a saline solution that he told them might relieve their pain. Brain scans showed that their brains produced more endorphins after they received the placebo. They also produced endorphins in more regions of the brain. Those men who said ahead of time that they expected the most relief had more endorphins than their more skeptical peers. The endorphin difference was pronounced in the brain's dorsolateral prefrontal cortex, a brain region involved in paying attention and making decisions.[18] It's as if the men made up their minds to get well and their brains cooperated by forming and releasing the proper medication.

Brain researchers such as Zubieta are working to sort through the complexity of the numerous brain regions and neurotransmitters that produce placebo or nocebo results. Theirs is no easy task. The placebo effect is not a single phenomenon, but the result of the complex interplay of anatomical, biochemical, and psychological factors. The same can be said for all the senses, I suspect. We see, hear, taste, touch, and smell pretty much what we expect to.

Chapter 5

Nematodes, Haptics, and Brain–Machine Interfaces

If you want to learn how the sense of touch works, ask a nematode. Granted, there's nothing particularly appealing about blind roundworms that slither around in the dirt all day, but nematodes are naturals when it comes to touch. Researchers have been studying them since the 1980s, and these creepy-crawlers have revealed a lot about how touch works in all animals, including people.

The nematode of choice is the lowly *Caenorhabditis elegans*, or *C. elegans* for short. At no more than a millimeter long, *C. elegans* isn't much to look at and it leads a dull life—by human standards anyway. It has no eyes, so its main sensory inputs are chemical and mechanical. Its days are all the same. It wiggles its way through the soil feeling the dirt, the bacteria it eats, and other nematodes—with whom it interacts only to reproduce—controlling all its sensory and motor activity through the precisely 302 neurons that make up its entire nervous system. About thirty of those neurons, roughly 10 percent, are devoted exclusively to touch.[1]

C. elegans is the darling of touch researchers because it's small and easy to maintain. It reproduces rapidly, lives its entire life in two weeks, and produces two hundred offspring, so there's never a shortage of research subjects. Most important, *C. elegans* has big, easily accessible neurons—each of which can be studied individually.[2] What's more, its mechanoreceptors (touch-sensitive neurons and their associated cells) are "conserved across evolutionary history." That means your touch sense works pretty much the same as a nematode's. Sorry if you're offended, but it's all a matter of chemistry, and once nature finds a successful mechanism, she sticks with it.

HOW TOUCH WORKS

The basic idea for any sense is the same: A stimulus prompts a neuron to generate an electrical impulse. In touch, the stimulus is mechanical. Pressure on the nerve endings embedded in mechanoreceptors such as Pacinian and Meissner's corpuscles (see sidebar) triggers an electrical signal. How? The electrical signal begins when a mechanical stimulus causes an ion channel in the membrane of a neuron to open. Ions are charged particles, and the channels are just what they sound like—tiny canals through which charged particles can enter a cell when a gate opens.

From studies of *C. elegans*, researchers know that a mechanical stimulus changes the structure of channels in the membrane of a touch-sensitive neuron. The change opens the channels' gates, allowing positively charged sodium (Na^{1+}) ions to flood in. The influx of ions changes the relative charges of the cell inside and outside, opening the gate to yet another channel, one that permits calcium ions (Ca^{2+}) to enter. That channel bears the unassuming name EGL-19. Activation of the EGL-19 channel sends an electrical impulse from a touch-sensitive neuron to an interneuron (one that carries messages between neurons). In the nematode, the interneuron ferries the signal to a motor neuron, which, in turn, stimulates a tiny roundworm muscle to contract, causing the animal to move. Signals from touch neurons also travel to opposing motor neurons that inhibit motion, so the animal can move in one direction, with movement in the opposite direction blocked.

And move this animal does, with a sophisticated repertoire of touch-triggered movements at its command. When it encounters an obstacle, for example, the worm reverses direction. The mechanoreceptors that produce that response are called ASH, FLP, and OLQ, among others. ASH neurons are also at work when the nematode turns away from poisons such as heavy metals, alcohol, and detergents.

When something touches its nose, *C. elegans* turns its head away from the stimulus. The mechanoreceptors OLQ and IL1 produce this head-withdrawal response. These neurons are also active when the roundworm is bobbing its head from side to side as it wiggles through the soil, hunting for food.

Researchers have done some weird things to *C. elegans*. One of them is to drag an eyelash hair along the animal's body to see if the worm responds to fine touch. It does, by reversing direction to move away from the hair. Exactly five neurons give the worm this sensation of gentle touch. (A sixth, while probably not required for gentle touch, is assumed to be a touch neuron because its structure is virtually identical to the other five. It may be a stretch receptor.) These touch neurons are arranged in a line along the worm's body, close to the skin's surface.

One way to find out how touch neurons work is to breed mutant roundworms and screen them for their loss of the gentle-touch response. Martin Chal-

A Historical Lexicon of Mechanoreceptors

In humans, touch is handled by neurons in the skin, joints, and mesenteries (membranes that support organs). Those neurons respond specifically and exclusively to a particular kind of touch, such as light contact, pressure, vibration, or body position. Some are free nerve endings that respond only to pain. Some are thermoreceptors; they initiate an impulse only when activated by heat or cold.

Others are specialized into classes of mechanoreceptors that vary in both how they are built and what they do. Mechanoreceptors are basically layers of cells organized in a distinctive way, with a mechanically responsive neuron nestled inside. In humans, the four main types are named after their discoverers:

- Meissner's corpuscles are sensitive to light touch. They are also important to the perception of texture, and they are most receptive to stretching. They occur near the skin's surface. They are named for Georg Meissner, the German anatomist and physiologist (1829–1905) who also has his name on Meissner's plexus, the network of nerve fibers that surrounds the small intestine.
- Merkel's corpuscles, also called Merkel cell-neurite complexes or Merkel endings, detect sustained pressure and low-frequency vibrations. They distinguish very fine texture changes, such as when reading Braille. Like Meissner's corpuscles, they are located close to the skin's surface. In 1875, Friedrich Merkel (1845–1919) discovered them in the skin of a pig's snout.
- Ruffini's corpuscles lie deeper. They detect sustained pressure. Also called Ruffini's brushes or Ruffini's end organs, they are named for Angelo Ruffini (1864–1929), an Italian histologist and embryologist who discovered them in 1894.
- Pacinian corpuscles detect deep pressure and high-frequency vibrations. Along with Ruffini's corpuscles, they lie deeper under the skin. They'll fire in response to a sharp poke, but not a sustained push. Their discoverer was the Italian anatomist Filippo Pacini (1812–1883), who reported them in 1840. He's also credited as the first to isolate the bacteria that cause cholera (1854).

fie and his team at Columbia have done just that, identifying about twenty genes named *mec* (mechanosensory abnormal). An abnormality in any one of the *mec* genes results in the loss of the gentle-touch response, without the loss of other normal movement patterns.[3] Each of the *mec* genes directs the cell to make a MEC protein, and each protein plays its own special role—either in the development of the touch neurons in the egg, larval, or juvenile stages, or in the operation of the mechanoreceptor in the adult worm. For example, if gene *mec-1* is abnormal, so is the protein MEC-1, and touch neurons don't attach to the worm's skin properly during larval development.

Another example is the gene *mec-4* and its protein MEC-4. That protein conducts sodium and perhaps calcium ions into the cell. If it's abnormal, ions don't rush in and the animal is touch insensitive, although the touch receptor neurons are structurally normal.[4] We know this because of a research technique that uses fluorescent dyes of different types. When apart, the dyes are invisible, but when they unite, they give off a fluorescent glow that can be detected through a microscopic technique called fluorescence resonance energy transfer (FRET). When the dyes are applied to touch neurons in *C. elegans*, low calcium concentrations keep the dyes apart and they remain colorless; but rising levels of calcium ions (Ca^{2+}) bring the dyes together to generate a greenish glow that FRET can pick up.[5] Such experiments have revealed that, among other things, MEC-4 and MEC-10 probably aid in pulling calcium channels open when the roundworm is gently touched.[6] Another likely candidate for that role is the OSM-9 protein, which belongs to a different molecular family, one called TRP.[7]

HAPTICS

The biochemical machinery of touch in *C. elegans* can keep molecular and cellular biologists enraptured for decades, but the engineers quickly lose interest. They have bigger bridges to build. A small but growing cadre of technological whiz kids wants to understand touch not as an end in itself, but as means to a number of other ends. These researchers want to know how the sense of touch works so they can replicate it in their machines, ranging from prosthetics for paraplegics to robotic technicians that can clean your teeth.

I went to see one such engineer at the Massachusetts Institute of Technology. He's Mandayam Srinivasan, director of MIT's Laboratory for Human and Machine Haptics, better known as the Touch Lab. *Haptics,* he says, is a broad term that encompasses both the sense of touch and the manipulation of objects by humans, by machines, or by a combination of the two in real or virtual environments. The researchers in Srinivasan's lab work on three fronts: (1) human haptics, the study of human sensing and manipulation through touch; (2) machine haptics, the design, construction, and use of machines to replace or augment human touch; and (3) computer haptics, the development of algorithms and software to generate and transmit a reality-simulating touch and feel from simulated, virtual objects—the computer graphics of the touch world, so to speak.

Srinivasan entered neuroscience research through the back door. He trained as both a civil engineer and an aerospace engineer; he earned his Ph.D. in applied mechanics at Yale. "My thesis had hundreds of equations, but no experiments. I was a totally theoretical engineer, an applied mathematician. . . . After my Ph.D., I got tired of shuffling equations, so I got into a neuroscience lab at Yale Medical School," he says. Srinivasan began to study biology there, but he retained his engineer's point of view: "I had expertise in the mechanics of deformable media," he explains, "and it turns out that mechanics is to touch what optics is to vision and acoustics is to hearing."

Srinivasan's engineering perspective paid off, for he was able to explore the territory that the biologists had not. To him, touch is a force applied against a semiflexible surface, the skin. In order to create the sensation of touch, without the bother of interacting with a real object, haptic engineers need only generate the proper kind of targeted force. At the haptic device interface, which can be something like a pen-shaped stick or a strip of pressure-actuators worn on the neck or arm, the human operating in a virtual world of touch receives signals generated via a computer. The signals create the illusion of hard–soft, rough–smooth, strong–weak, vibrating–still, or any other touch perception the haptic engineer wishes to generate.

To Srinivasan, nature's touch interface—the fingerpad—is a complex of interactive tissues exhibiting various forms of mechanical behavior, including inhomogeneity (composed of different parts) and anisotropy (unequal physical properties along different axes). The various sensory detectors in the fingerpad also operate at different rates and exhibit varying degrees of time dependence in their functions. Skin varies in its compliance (how well it conforms to the force applied) and its frictional properties (responses to various textures). The mechanical properties of skin and the underlying tissues determine how forces "load" on its surface, how they are transmitted through its layers, and how the forces are detected and signaled by skin's near-surface and deeper-down mechanoreceptors.[8] To Srinivasan and his team, identifying those factors leads in two directions: Write a series of mathematical equations to explain how touch works and use those equations to make machines deliver touch sensations, either to a human being or to a robot.

In the MIT Touch Lab, Srinivasan's team has developed both the hardware and the software that humans need to interact with objects that aren't there. Perhaps the most famous is the PHANTOM, a deceptively simple pen-shaped probe now marketed by SensAble Technologies (Woburn, MA). The PHANTOM gives its user a sense of the shape, softness, texture, and friction of solid surfaces that don't exist.

The potential uses of haptic devices such as the PHANTOM are legion. One big area is the training of medical personnel. Just as flight simulators train pilots, surgical simulators let medical students see, touch, and manipulate convincing models of tissues and organs long before they cut into the real thing. Srinivasan's lab has developed a simulator for training anesthesiologists to give epidural injections. A minimally invasive surgery simulator is being developed, too.

Srinivasan made headlines in 2002 when he collaborated with haptic researchers at the University College London to demonstrate the first instance of "transatlantic touch." Two users, working via the Internet in London and Boston, collaborated in lifting a virtual-reality box. Using the PHANTOM to receive the sensation of touch, each lifter could feel the force of the box and experience the changes in its position as a distant collaborator acted on it.[9]

Other potential applications of haptic research include the following:

- *Medicine:* micro- and macro-robots for minimally invasive surgery; remote diagnosis for telemedicine; aids for the disabled, such as haptic interfaces for the blind
- *Entertainment:* video games and simulators that enable the user to feel and manipulate virtual solids, fluids, tools, and avatars
- *Industry:* haptic-feedback, computer-assisted design systems that allow the designer to manipulate the mechanical components of an assembly during the design stage
- *Arts and education:* concert rooms, virtual art exhibits, and museums in which the users can "play" musical instruments or "sculpt" their own statues[10]

BRAIN–MACHINE INTERFACES

As the deceptively simple prototypes I examined in Srinivasan's lab reveal, real-life haptic devices are nowhere near what Hollywood depicts. Haptic interface tools can transmit the sensation of force to the user, but their range, resolution, and frequency bandwidths have a long way to go before they can match the human sense of touch. To engineers like Srinivasan, such improvements will come from better calculations of mechanical forces and more "degrees of freedom" in robotic manipulators. (Degrees of freedom are the points at which movements are possible, such as—in the human arm—various independent and combined positions of the shoulder, elbow, and wrist.) Tactile simulators need the fine resolution that human skin can muster. Tactile arrays have other design constraints. They must be inexpensive, lightweight, comfortable, and compact enough to be worn without impeding the user's movement. There's also the problem of synchronizing visual and auditory information with haptic feedback—no small matter since the different senses work via different inputs and at different rates. "The current models of virtual objects that can be displayed haptically in

Why Lo-Tech Back Scratchers Feel High-Tech Good

It's not a high-tech gadget, but the old-fashioned back scratcher is one of the most effective of all haptic devices. Until recently, no one understood why scratching is so pleasant, but Wake Forest University researchers changed all that in 2008 when they disclosed what goes on in the brain when we scratch.

The study, led by dermatologist Gil Yosipovitch, reported fMRI scans of the brains of thirteen healthy participants while their lower legs were scratched with a small brush. Scratching periods of thirty seconds alternated with resting periods of the same length for a total of five minutes.

Yosipovitch found that areas of the brain associated with negative emotions and memories became less active during scratching. The reduced activity occurred in the brain's anterior cingulate cortex, an area associated with aversion to unpleasant sensory experiences, and the posterior cingulate cortex, which is associated with memory. When participants said that the scratching felt most intense, activation in these areas was lowest.

Some areas of the brain became more active during the scratching; they included the secondary somatosensory cortex, a sensory area involved in pain, and the prefrontal cortex, which is associated with compulsive behavior.[11]

"We know scratching is pleasurable, but we haven't known why. It's possible that scratching may suppress the emotional components of itch and bring about its relief," says Yosipovitch. The activation of the prefrontal cortex "could explain the compulsion to continue scratching," he adds.[12]

One drawback to the study is that the scratching occurred in the absence of a genuine itch. Yosipovitch's team is continuing the research to find out if the brain acts the same when that annoying itch just *has* to be scratched.

real-time are quite simplistic compared to the static and dynamic behavior of objects in the real world," Srinivasan says.[13]

One of the goals of haptic research is to develop devices that will assist people who have lost the use of a limb to accident or disease. In 2008 researchers at the University of Pittsburgh made progress toward that goal when two monkeys in their laboratory successfully used the power of their brains alone to control robotic arms and hands. Unlike similar, earlier experiments in which brain power controlled the movement of a cursor on a screen, this robotic action was three-dimensional and personal. The monkeys picked up food, fed themselves, and even licked their robotic fingers.

To achieve this level of brain–machine communication, the researchers first analyzed the collective firings of neurons in the animals' brains when the monkeys moved their natural arms. Then the investigators wrote software that translated the patterns of neuronal firing into signals that controlled the movements of the robotic limb. Although the monkeys received no sense of touch feedback per se, they were able to control their robotic arms in a natural way, and the feedback—successful food retrieval and feeding—was sufficient to sustain the action. So precise was the control that the monkeys could adjust the grip of the robotic hand to accommodate varying sizes and thicknesses of food pieces. To the researchers' surprise, the monkeys learned that marshmallows and grapes would stick to their hands, so a firm grip wasn't necessary when those foods were on offer.[14]

Such achievements in haptic research are laudable, but the field still has a long way to go. "I could talk for a year on what I have discovered and probably ten years about what I have not," Srinivasan says. He recalls a list of his top ten questions about touch he wrote when he first entered the neuroscience research lab at Yale Medical School in 1983. "I think I have probably barely answered about three of those ten, but at the same time I have answered ten more that were not on that list," he says. But each of those questions has spawned ten more. "We researchers are never out of business," Srinivasan says.

PART TWO
SMELL

Of our five senses, that of Smelling has been treated with comparative indifference. However, as knowledge progresses, the various faculties with which the Creator has thought proper in his wisdom to endow man will become developed, and the faculty of Smelling will meet with its share of tuition as well as Sight, Hearing, Touch, and Taste.

—G.W. Septimus Piesse
The Art of Perfumery
1857

Chapter 6

It's Valentine's Day . . . Sniff!

It's Valentine's Day. Close your eyes and open that heart-shaped box. How do you know there are chocolates inside? Why, that sweet smell is both unmistakable and unforgettable.

But what makes that smell unique? If you stop to think about it, there's a lot that goes on in detecting that odor, knowing it's chocolate (and not Brussels sprouts), remembering how much you love chocolate, and actually taking a piece.

Let's look at how the brain handles smells, step by step.

WIRED FOR SMELL

The sense of smell is hardwired into the human nose and brain. Molecules of odorant move up through the two nasal cavities until they come to the olfactory epithelium. This region in the nasal lining, about the size of a postage stamp, contains 10 to 20 million olfactory neurons. Spiny projections (dendrites) of these neurons extend to the epithelial surface where they end in knobs. On the knobs are tiny hairs that float in a layer of mucus. The hairs contain odor receptors, one kind per cell, among the thousand or so that humans have. Odor molecules fit receptors the same way that keys fit locks. When an odoriferous molecule binds to its perfectly matched receptor, a nerve impulse begins. (Or maybe there is more to it; see Chapter 7.)

Particular types of receptors are limited to specific regions of the olfactory epithelium, but thousands of cells express a single receptor. Most receptors respond to several different odorants, but some fit only one (or a few very similar ones).[1] When an odorant molecule binds to a receptor, the receptor releases a

The Chemistry of Smell

To activate the nose/brain smell machinery, a molecule must be:

- Volatile (easily evaporated into the air)
- Water soluble, to pass through the mucus inside the nostrils and reach the olfactory cells
- Capable of attaching to the surface of receptor cells

Many molecules meet those criteria, but few people agree on how odors are best described and classified. The eighteenth-century scientist Linnaeus proposed seven categories of smell: camphor, musk, floral, minty, pungent, putrid, and ether. More than a century later, the German psychologist Hans Henning had a different idea. He constructed a "smell prism" using six odors: flowery, fruity, spicy, foul, burnt, and resinous. He said every odor could be placed on a three-dimensional grid created by those six points.

Today, many other systems are in use, but none captures to everyone's liking the range and subtleties of what the nose "knows." One of the problems is that the size and shape of molecules should predict their odor, but they don't always. In general, chains of four to eight carbon atoms in aldehydes and alcohols have strong odors. The odors of six-carbon benzene rings depend on where side chains of carbon, hydrogen, or oxygen atoms are attached. Large rings of fourteen to nineteen carbon atoms can be rearranged a lot without changing their odor.

The way an odorant is perceived can even vary with its amount. For example, indole in small amounts smells like flowers. In high concentrations, it smells putrid.

G protein. The release of the G protein triggers another molecule—a "second messenger"—that causes a nerve impulse to begin. The impulse travels from the cell body of the neuron along its long, fibrous axon. Axons from many neurons merge into axon bundles and into the olfactory nerve (cranial nerve I). The cells that express a single receptor type send their axons into a single point in the olfactory bulb of the brain.[2] Those are points of "synaptic convergence" are called glomeruli. The convergence must let the brain know which type of receptor is being stimulated.

Unlike impulses from the other senses, odor signals are not sorted in the brain's central switching and relay station, the thalamus. Rather, they travel directly to the brain's center of emotion, the amygdala, and to its memory-forming center, the hippocampus. From the glomeruli, signals travel also to the primary sensory cortex, including regions of the frontal lobes. In right-handed people, the right hemisphere of the brain is slightly dominant over the left in processing smells.[3]

Once the transmission is received in various regions of the brain, appropriate messages are sent to other parts of the brain and body. Activity in the language and emotion regions of the brain's cortex lets you name and appreciate a smell ("Hmmm, chocolate!") and act on it ("I'll take a piece!")

A MATH PROBLEM

Although signals begin in a single type of receptor, many receptors bind to more than one type of odorant. Also, odorants activate more than one kind of receptor. So how does the brain end up distinguishing one smell from another? The answer lies in a series of connections among neurons that carry messages into several regions of the cortex.

In 1991 Columbia University scientists Linda Buck and Richard Axel discovered the receptor proteins that bind to odorant molecules. Buck and Axel received the Nobel Prize, and research on smell has boomed in the last twenty years, much of it directed toward the puzzling question of how humans can detect some ten thousand different odors using only one thousand different kinds of receptors. (To make the math problem worse, only about 350 of those receptors actually work. The others have lost their function over human evolutionary history.)

To solve the math problem, Japanese scientists worked with Buck to test smell receptors in mice. They loaded olfactory neurons with a fluorescent dye. If an odorant activated a neuron, the dye changed color. The researchers showed that different odorants activate different combinations of receptors. The brain identifies odors by evaluating patterns of impulses. To oversimplify for the sake of example, imagine that chocolate candy stimulates receptors A, B, and C. Simmering hot chocolate stimulates B, C, and D. The brain distinguishes the two because the combinations are different.

These complex and spread-out patterns create "odor maps" in the brain. In the olfactory cortex of the brain, a single odor is represented by activity in a distinct area of sparsely distributed neurons. Different odorants elicit different maps, but the maps can overlap. An increase in the concentration of an odorant expands the area of the map and triggers activity in additional neurons in the brain's cortex. Odorants that are chemically similar have maps that look a lot alike. Combinations of two odors activate neurons that neither odor alone can account for.[4] To Nobel laureate Linda Buck, such mapping patterns suggest "an underlying logic to the mapping of odor identities in the cortex."[5]

That logic and the odor maps that result from it are remarkably similar among individuals. Although experience may affect scent preferences, we are all hardwired for smell pretty much the same.[6] And just as the twenty-six letters of the alphabet can be combined in different ways to produce millions of words, the impulse patterns from 350 receptor types can combine in multiple ways to allow humans to recognize some ten thousand different odors. Cornell University scientist Jane Friedrich explains the process this way: "Let's say you smell an essence oil like jasmine, which is a pure smell. Your ability to smell that jasmine is based on a small combination of olfactory receptor proteins. Those proteins produce a pattern your brain recognizes."[7]

READY, SMELL, GO!

While we often assume that our reactions to smells are inborn—for example, babies smile when they smell vanilla and cry when they smell rotting meat—we may not be as hardwired for scent preferences as we think we are. Researchers at Rockefeller University studied a single neuron in the olfactory system of the nematode *C. elegans* (see Chapter 5). That neuron is especially sensitive to the odor of butanone. The scientists found that nematodes that were fed when the scent of butanone was in the air quickly learned to love the fragrance. They wiggled and squirmed their tiny bodies eagerly toward the odor. But if the air smelled of butanone when the animals were starved (for only two hours; these researcher were not cruel to nematodes!), the animals later ran away from the butanone scent. The researchers demonstrated that experience altered the neuron's response to the scent, triggering a change from attraction to avoidance. People aren't nematodes, and our olfactory systems work differently, but the Rockefeller team suspects olfactory neurons in the human brain may show the same adaptability.[8]

That research may help explain why individuals vary in what they smell and in how we perceive an odor. In a classic experiment performed over thirty years

Smell Cells of a Different Breed

Until recently most scientists thought that only olfactory sensory neurons detect odors. They were wrong. In the upper reaches of the nose lies a different category of cells, the solitary chemosensory cells (CSCs). CSCs aren't your everyday odor detectors; they send their messages only when the concentrations of an odorant are extremely high. The message is more than one of increased intensity. It's the noxious, irritating message of a smell that can cause burning, stinging, itching, or pain. CSCs send messages to the brain via the trigeminal nerve (cranial nerve V). That nerve usually carries impulses of touch and pain from the face.[9]

Take Time to Smell . . . (the Mints)

Sniff and sniff again. Taking a tenth of a second longer to savor a scent increases your chances of identifying it, say researchers at the Monell Chemical Senses Center in Philadelphia.[10] That observation is of more than passing interest. It reveals a lot about how the brain processes olfactory inputs. As Buck and Axel showed, odorants stimulate several different receptors, and the brain responds not to a single signal but to a pattern of multiple signals. Some kind of code must lead to identification, but its content remains uncertain.

Jay Gottfried at Northwestern University has taken one step toward cracking that code. He's shown that learning smells requires no effort. Simple exposure is sufficient, and the longer the time, the better. He used fMRI to record brain activity in people exposed to minty and floral scents. The subjects sniffed the odors for more than three minutes. The lengthy exposure taught them to discriminate not just a single minty or floral odor, but numerous variants within each class.[11] "When you have prolonged sensory experience with one smell, you become an expert for smells that are part of that original category," says Gottfried.[12] That expertise revealed itself in the brain scans. The longer a subject sniffed, the greater the activity in the brain's orbitofrontal cortex, a region associated with emotion and motivation as well as smell.

ago, test subjects sniffed isobutyric acid. Most of them said the acid smelled like dirty socks, but a few said it had a pleasant, fruity odor. It turned out that those few weren't smelling the acid at all. Their noses and brains couldn't detect it. The fruity smell was coming from impurities in the acid. Such experiments suggest that some of us are nosy superstars, while others are scent simpletons. The difference may not matter much in everyday life, but it matters a lot when manufacturers hire panels to judge the scent of a new perfume or the aroma of a Valentine's Day treat.

Individual differences aren't the only mysteries waiting to be solved. Scientists are also trying to learn why we notice odors in the first place and why we stop smelling them if they linger. Johns Hopkins researchers say that a nerve cell "chorus" may enable the brain to pay attention to one thing above all others. Neurons firing in unison signal *Attention!*—allowing you to home in on chocolate and ignore everything else.[13] The brain's motor centers get in the act, too. An unconscious message to the nose says, *Sniff!* Sniffing brings a rush of air—and more of the odorant—into the nose. Stanford scientists say that sniffing acts as a wake-up call for the brain. It primes the brain to receive a smell impulse.[14]

Other scientists, also at Johns Hopkins, have found what they believe to be the "nose plug" that makes us stop smelling an odor that lasts a long time. It is actually a channel in the sensory neurons of the nose that opens when an odor is present and closes as the neuron becomes desensitized to the smell. A protein called CNGA4 does the job.[15]

There's no doubt that odors can be learned, but how that learning occurs remained a mystery until 2009, when researchers at Case Western Reserve University discovered "synaptic memory" in the brain's smell centers. Usually, neuroscientists investigate how impulses move from lower down in the brain (where impulses arrive) to higher up (where impulses are interpreted), but Ben Strowbridge and his team traced a pathway running in the opposite direction, through a rarely studied projection of neurons that runs from the brain's olfactory cortex back down to the olfactory bulb. Strowbridge found evidence of long-term potentiation (LTP), which is the strengthening of the connections among neurons the more they are used. LTP is the basis of all learning and it's known to occur in virtually all areas of the brain, but its occurrence in the olfactory system had never before been demonstrated. Based on his studies, Strowbridge thinks that brain regions do a lot of talking back and forth when smells are learned.[16]

The processes that explain the learning of smells may also explain, in part, why odors are powerful triggers for behavior. Contrary to conventional wisdom, the human sense of smell is remarkably acute. A research team led by scientists at the University of California at Berkeley asked blindfolded volunteers to follow a trail of chocolate scent merely by sniffing. Most people accomplished this task with ease, and their performance improved with practice.[17] What's more, their brains automatically associated the sweet smell of chocolate with the reward chemicals the brain makes in response to the taste of a gooey, fudgey goodie.

It's enough to make you pause for thought, before helping yourself to another piece of Valentine's Day candy!

Chapter 7

Do Odors Dance in Your Nose?

While I'll admit to an occasional daydream about Denzel Washington or Jimmy Smits, the true object of my unbridled imagination is Luca Turin.

Here's how my daydream goes.

I slave over this book for well over a year, and after it is published, it becomes a bestseller overnight. The world is beating a path to my door, but I'm unwaveringly modest and retiring until I get a call from Luca Turin. He's read my chapters on smell and loves every syllable, so he invites me to London to visit and discuss his one and true love—perfume.

Even the writers of bestsellers have to wait six months for their royalties, so I decline the invitation because everyone knows about the price of fish and chips in England these days. But Turin insists, reassuring me that he has already persuaded one of the Big Six perfume manufacturers to foot the entire bill, including a first-class seat for me on British Airways and a deluxe suite at the Dorchester.

Well, I can hardly turn down a deal like that, so I'm on the next flight, having purchased a brand new bottle of Sécrétions Magnifiques for the trip. Turin sang its praises in a perfume review as a brilliant amalgam of "harbor bilge in an elegant floral."[1] That sounds like the signature scent for me, so off I go with many of my fellow passengers on the airplane keeping a respectful distance. Could it be the harbor bilge?

Never mind. In London, Turin greets me enthusiastically and ushers me into his creatively messy office. I know what to expect because I read *The Emperor of Scent*, Chandler Burr's book about Turin:

> The office . . . looked like a hand-grenade test site. Transistors, wires, tubes, plane tickets, bottles of perfume, obscure scientific journals and copies of *Vogue* and magazines about airplanes, gadgets of every size and design and God knows what else, and everywhere, vials and vials and vials.[2]

In my daydream, Turin begins extracting vials from those jumbled piles, opening them, waving them under my nose, explaining to me the brilliance of a touch of indole (flowers) in this one and a touch of skatole (feces) in that one! In my daydream, I'm no different than in real life—I can't tell *muguet* from mop water—but Turin has me totally captivated. He's animated, charming, eccentric, passionate—Russell Crowe in *A Beautiful Mind* without the schizophrenia. He explains everything there is to know about smell and scent, and I soak it all up like a scratch-and-sniff patch.

But who, you may be wondering, is Luca Turin and why do I daydream about him? To my knowledge, he's never been on the cover of any magazine. He's never made a movie, not even a commercial. He's never received a Grammy or an Oscar or a Tony, although he's been described as a sharp dresser, so I can imagine him on the red carpet. He's said to be a "polymath, a brilliant, intellectually restless, self-educated aesthete."[3]

That's all fine, but the main reason he's my idol is that he's bent some very prestigious academic noses out of joint with his highly refuted theory about how the sense of smell works. To my mind, the only thing more fun than science is controversy, and Turin is swirling in a skatole-scented maelstrom of both. He's brazenly and unabashedly the maverick of scent science, and that sparks my imagination. Here's the essence.

LUCA TURIN AND VIBRATION THEORY

Conventional wisdom holds that molecules of odorants lock onto receptors in the nose. They fit there because of their size and shape. When they dock—like a boat in a narrow slip—they cause a subtle change in the shape of the receptor. That shape-change triggers a nerve impulse. Next thing you know, you smell burned toast.

Luca Turin thinks that's wrong. He says shape is only one part of the explanation for how we smell. He says that smell receptors in the nose respond to the *vibration* of molecules after the docking. Perceived odors differ because the *frequency* of molecules' vibration differs—not just their size and shape. To support his assertion, Turin points out that some molecules that are almost identical in shape smell very different. Others that have different shapes, but vibrate at the same frequency, smell more or less alike.

Turin isn't the first to propose such an idea. In the 1920s and 1930s Malcolm Dyson, who had the misfortune to be gassed in World War I, experimented (perhaps predictably) with the class of compounds called isothiocyanates: mustard gas and phosgene. He started adding atoms to the compounds and changing their shape. Some of his compounds smelled like mustard. Others smelled liked anise. When he started to compare his molecules' architecture to their smell, he found disparities. Some of his molecules had similar shapes, but they smelled different, while some with different shapes smelled the same. Dyson knew that molecules vibrate, but he had no way to measure their frequencies. Nonetheless, he proposed that the human nose must be responding to "osmic frequencies."

No one paid much attention to Dyson's idea, and those who did punched holes in it, but in the 1950s Canadian scientist Robert Wright breathed new life into the notion. He proposed that (in Luca Turin's words) "some receptor would actually feel the molecule buzz."[4] Wright made some good arguments, and he continued espousing them through the 1970s. Unfortunately, the range of "smellable" frequencies he postulated was too narrow, and his ideas never garnered much serious attention.

Then, in the 1990s, along came Turin. He suggested that the vibrational frequency of the molecule offered a better explanation for odor than did simple molecular shape.[5] Turin rejected the idea that each type of smell receptor might respond to a different frequency. He asserted instead that each type responds to the same frequency *range*. Some molecules have no odor because they are too large to fit into a receptor. Small molecules that vibrate but have no smell could, Turin said, lie below the threshold of frequencies that stimulate olfactory receptors.

GOTCHAS!

The skatole hit the atomizer when Turin tried to get his hypothesis published in the prestigious British journal *Nature*. His paper was peer reviewed by experts on smell and summarily dismissed as nothing short of nonsense. Shape theory explained all that needed explaining, and Turin's theory had more holes in it than a butterfly net, the journal editors concluded.

But a skatole-storm started brewing when the press got a whiff of unrest in the scientific ranks. Turin's opponents were forced—in public and in private—to throw some gotchas his way, and there was no shortage of dissent. "Simply because two different molecules may interact with different receptors does not necessarily mean that they will smell different," Tufts neuroscientist Joel White told Stephen Hill of the *NewScientist*.[6] Differing *combinations* of receptor activation could account for similar or dissimilar scents—once the brain has its say, White said. Others concurred. "The brain puts together an odour based on the pattern of signals," said perfume expert Charles Sell.[7]

Turin countered that gotcha with some gotchas of his own. If different signals to the brain could produce the same perception of odor, he reasoned, then could the opposite possibly be true? Could a single set of receptors lead to the perception of two different odors? Turin argued the answer was no if vibration explained the sense of smell, but it could happen if size and shape were the only variables. He offered the compounds ferrocene and nickelocene as examples. They are the same size and shape, but ferrocene has an iron atom in it. Nickelocene has a nickel atom. Nickelocene has hardly any odor; ferrocene smells spicy. Turin said vibration explains the difference. Nickelocene has a low-frequency vibration, below wave number 400, which Turin says is the threshold of scent detection. Ferrocene has a much higher number, approaching 500, so receptors in the nose respond to it.

Another big gotcha was the enantiomers. They are mirror-image molecules. They have the same structure, but they are flipped right to left; they have the same vibration, but a different shape. Admittedly, most enantiomers smell the same. In one category, however, the enantiomers smell different. They are the carvones. R-carvone smells like spearmint. S-carvone smells like caraway. Go explain that, Turin!

Turin retaliated. He said that the vibration of the different-smelling carvones is the same only in the solution or gas phase used to test such things in the laboratory. He said smell receptors twist the molecules differently and dock with different segments of them. That means that the docking sites vibrate at different frequencies. To prove his point, Turin mixed the minty-smelling carvone with another compound that changed the molecule's vibration but not is shape. The result? The smell of caraway.

Still another of Turin's repudiations involved hydrogen. Hydrogen atoms exist in three isotopic forms. The ordinary hydrogen atom has one proton in its nucleus and one electron in its orbit. Electrons weigh too little to worry about, so the proton accounts for most of hydrogen's mass. Another isotope of hydrogen is deuterium. It's the same as ordinary hydrogen except it has an uncharged particle, a neutron, added to its nucleus. Because both have a single electron in orbit, their size and shape are the same, but their mass is different. So when they are added to big, organic molecules, they don't change the big molecule's shape and size, but they do change its vibrational frequency. As a rule, the heavier the molecule, the slower its vibration. Guess what! When Canadian chemists used deuterium to replace all the ordinary hydrogens in a certain category of compounds called benzaldehydes, the odor was different.[8] Vibration theory could explain that. Shape theory couldn't.

The Holy Grail, in Turin's view, was the search for two molecules with different shapes, but the same vibration, that smell the same. The search was long and difficult, but eventually Turin found a boron compound, decaborane, that smells like rotten eggs (hydrogen sulfide) although it has no sulfur in it. Its shape can't explain its sulfur smell. Its vibrational frequency can.

SHOWDOWN

After Chandler Burr's *The Emperor of Scent* was published, lots more gotchas were flying. Some people loved the book; others deplored it. As for Turin's theory, one reviewer concluded:

> While implausible, Turin's proposal is certainly a delightful potpourri of creativity, conjecture, extrapolation, and isolated observations. And it's brazen: a universal theory of smell based on one man's olfactory perceptions. In a grand substitution of ego for psychophysics, Turin claims that Turin's theory successfully predicts odors because they smell the way Turin says they do.[9]

In 2004 Rockefeller University's Andreas Keller and Leslie Vosshall tried to settle the matter for good by doing what scientists do—testing the hypothesis.[10] They asked volunteers to smell different odors from vials; the vials were numbered so that neither the researchers nor the subjects could tell what was in them. The research subjects then reported what they smelled.

In one experiment, the researcher mixed two chemicals. One smelled smoky. The other smelled like bitter almonds. Turin's theory predicted that a mixture of the two should smell like vanilla. It didn't. In another experiment, the Rockefeller team worked with a class of compounds called the aldehydes. Vibration theory predicts that aldehydes containing an even number of carbon atoms should smell something alike (because their vibrational frequencies are similar). Their odors should be noticeably different from aldehydes with odd numbers of carbon atoms and a different vibrational frequency. Again, no foul-smelling cigar. The volunteers didn't find the scents of two odd or two even aldehydes any more alike than an odd and an even. Finally, Keller and Vosshall tested Turin's hydrogen and deuterium compounds, criticizing the previous Canadian work on methodological grounds. In the Rockefeller study, participants couldn't tell the difference between deuterated and nondeuterated aldehydes.

In an editorial that accompanied the Rockefeller paper, *Nature*'s editors pulled no punches. They were publishing a paper testing Turin's hypothesis not because the notion had merit, they said, but because it had received so much publicity from "uncritical journalists." Apparently, the editors didn't like Turin any more than they liked the media:

> A mature body of scientific theory is like a large building, and the impulse to demolish it is often little more than a form of intellectual vandalism,

The Problem with Peer Review

The experts started vibrating over vibration theory when Turin submitted a paper about his idea to the prestigious British journal *Nature.* The journal rejected the paper. Turin's ideas couldn't stand up to peer review.

Now let's stop talking about smell for a minute and turn our noses toward the slightly odiferous process of scientific peer review. Peer review is a lot like democracy; it's a lousy system and the best we've come up with yet. On the face of it, it makes sense. Let's say you're an expert on the Patagonian raptor *Milvago chimango.* You've been out in the field studying the bird and you've written a paper on the birds' nesting habits. You are having hot flashes of eagerness to publish your paper, but it will never see print unless it is favorably reviewed by other experts on the *Milvago chimango.* Their opinions, fair or foul, will determine your fate. Surely, that's wiser than having reviewers who wouldn't know a raptor from a lab rat judging your work. It makes sense: like evaluates like.

In the best of all possible scientific worlds, your reviewers are professionals who respect you and your work. They are as passionate about Patagonian raptors as you are, and they have seen for themselves, during lowland treks through Argentina and Chile, the same nesting habits you have seen. Heaps of grant dollars are lying around; no one is competing for funds. Your reviewers give your paper an A+ and *ka-bam!* You get your five minutes of fame on the pages of an erudite journal.

But suppose your alleged peers never heard of you (and if they have, they judge you to be an eccentric). You've been working on the Chilean side of the mountains, and they have been working on the Argentinean. Suppose, also, that every paper your reviewers have ever published says that *Milvago chimango* nests high in trees. They've seen the birds there. They've observed thousands of nests. But you have been beating the bushes, and now you have elegant analysis of variance tables and histograms to counter the prevailing wisdom. Most of the birds pick lower nesting sites, you have discovered, at or near the center of their "substrate," whether it be a tree, shrub, cliff, or pole. You've even documented a small, but significant, number of nests on or near the ground.

So, how are your peer reviewers going to react? They never heard of you and your idea is ridiculous. Why should you get grant funds? Case closed.

Now, I'm not saying that is what happened to Turin. Well, maybe I am, because Turin is, indeed, looking on a different part of the tree and spotting a different kind of nest. His peers have not seen it, and they aren't inclined to try. All their research is built on another "substrate," and who's going to ditch a lifetime of work (plus a solid reputation) to jump for some bird-brained idea out of nowhere?

Back to Turin's book. *The Secret of Scent* isn't about Turin. It's about science, written large and painted with a wide brush. Turin trained as a biologist, but he refuses to stay focused on the life-science branch of the tree. He incorporates anatomy, physiology, chemistry, physics, and a good deal of technology into his vibration theory. His eclecticism sets him at odds with peer reviewers. Specialists specialize. They limit their vision to a single phenomenon, narrowly and tightly defined. There aren't many who look—as Turin does—at the whole tree, bend its branches, and find nests where no one else spotted them before.

Or maybe he's just sniffing up the wrong tree.

> an expression of frustration by those who did not succeed as architects. Some buildings outlive their usefulness, of course, but the threshold for knocking them down should be high. We hope that the paper from Keller and Vosshall will serve as a reminder of why that's so.[11]

To this day, Turin stands firm. He's right, he insists, and he emailed me some answers of his own, rebutting the Keller and Vosshall studies point by point. Then, he refused to give me permission to publish his rebuttals. I can only guess why. Distrust of an "uncritical journalist"? I don't think so. We're the ones who listen to him when others won't. Lack of confidence in his own ideas? Given the nature of the man's public character, highly doubtful. Eccentricity? Perversity? Who knows. Maybe he's tired of beating a dead horse. Maybe he's tired of bellowing at one.

So, for now, let the record stand. In the *Secret of Scent*, Turin devotes less than two pages to the Keller and Vosshall refutation. He acknowledges that isotope effects are subtle and might not be easily discernible and then dismisses the Rockefeller study as "a valuable data point in a continuing debate."[12] He marches on without a sniff of remorse, reporting how his theory gives perfumers a one in ten chance of finding a fragrant molecule. Shape theory, he contends, gives a one in one thousand chance.

He may be right about the improved odds. In 2004 Cambridge researchers organized forty-seven molecules into seven odor categories: ambergris, bitter almond, camphor, rose, jasmine, muguet, and musk. Then they assessed both shape and vibration to see which better predicted a compound's fragrance. Vibration won.[13] Although the researchers carefully avoided choosing sides in the dispute, they didn't deny that their findings have practical advantages for chemists in the scent industry.

Recently, Turin's hypothesis gained additional support—so much so that even *Nature* chalked up a gotcha for Turin.[14] Some colleagues of Turin's at the University College London created a mathematical model of how the vibrational interaction between a receptor and an odorant molecule might work.[15] In the model, the receptor has two "wells" separated by an energy barrier. One of the two wells contains a bound electron that can break loose and tunnel through the barrier. When the odor molecule binds to the energy barrier site, the electron goes tunneling though it, setting the molecule to vibrating—something like a pick plucking a guitar string. That triggers a nerve impulse that goes zipping off to the brain.

Electron tunneling is an intriguing idea, and it explains some facts about smell, such as how rapidly odors are detected in the brain—in less than a millisecond, the experts who measure such things say.[16] Still, the London-based scientists only worked out the mathematics of it; no one has yet observed leaping electrons in anybody's nose.[17]

Perhaps by the time I take my dream trip to London, someone will have.

Chapter 8

Life Without Scent

Ask a hundred people on the street which of their five senses they'd be most willing to give up, and most will answer smell. Yet smell is vital to human health and survival. It alerts us to danger and disease. Odor often tells us which foods are safe to eat and which are spoiled. Scent attracts us to friends and lovers and sends us running when fire, skunks, and chemical spills threaten. Much of the flavor of food results from our sense of smell, so the pleasure of eating is impaired in those who lose their olfactory sense—a condition called anosmia.

Theresa Dunlop knows what anosmia is like. She was injured in an accident in 1974, when she was twelve. "I was hit by a car and flew twenty feet in the air and landed on my head. I broke my femur bone and wound up in traction and a body cast. The lump on my head was so bad that my parents would not let me see myself in a mirror for a month," Theresa recalls.

No one said anything about the possible sensory effects of the head injury, but it wasn't long before Theresa realized she had lost her sense of smell. She couldn't remember the smell of her grandma's cookies or decaying leaves in autumn. She couldn't recognize the odor of dirty laundry or a burned dinner.

Her anosmia proved hazardous. "Since I was in traction the nurses did not want to mess with my bedpans at night so they cut me off from drinking by a certain time. One night a nurse brought me nail polish remover in a little cup. When she got buzzed and went to help someone else, I saw the cup and thought, *Oh, yay! She forgot and left me some water.* I shot the polish remover in my mouth quickly. I never tasted it on my tongue, and it wasn't until it was burning my throat that I realized what a dumb thing I had done. I gagged and spit it up."

The doctors and nurses blamed the incident on Theresa's poor memory. No one mentioned her sense of smell. It was years before Theresa understood the source of her anosmia. "When I was a young adult and going to hairdressing school, I had to study the bones, nerves, and muscles of the face and head. It was then that I realized that the scar on my head was the part of the brain that controls the sense of smell."

Theresa says her anosmia has continued to trouble her in adult life. She had a baby and couldn't tell when a diaper needed changing. She was spending so much time peeking into the diaper that she finally established a schedule, changing the baby by the clock. Theresa has accidentally mixed noxious bathroom cleaners, growing faint and dizzy as a result. For a long time, she worked as a hairdresser, but she couldn't smell the fumes of the dyes and perms. The chemicals gave her headaches, although she was unaware of the cause. "I would not sense the need to step aside from many smells," she says. "I would often get nauseous because the smells seemed to bypass my nose and go straight to my stomach."

Theresa grew compulsive about cleanliness. She brushed her teeth too often because she couldn't tell if she had bad breath. She marks on a calendar the showers her teen sons take; she worries that one of the boys might have body odor and she won't know. Theresa doesn't wear fragrances. "Once in my life I might have wanted to hear, 'Hmmm, you smell good,' but the one time I tried to use perfume, I used too much and gagged my family," she says.

The kitchen poses a special challenge to an anosmic like Theresa, especially because smell contributes so much to the sense of taste. For a long time, Theresa consumed several bottles of ketchup weekly and lots of vinegar, too, "just to feel like I tasted something," she recalls. "It was very hard to learn to cook. I had to follow recipes closely. Now, I don't hog the ketchup bottle anymore, but I do have tomato sauce several times a week." To prevent burned dinners, she relies on family members and smoke detectors in every room. She can't have a gas stove; how could she tell if the pilot light went out?

Human beings are remarkably adaptable, and Theresa is no exception. Most of the time her brain lets her forget about her anosmia, but sometimes her body reminds her of what she's lost. "I can smell things one time a month. I think it is when I am ovulating—no kidding! Once in awhile I say, 'What's that smell?' and those I love grimace and say, 'Oh, no,' because they know PMS is on its way. . . . On certain days when I get a brief quick whiff of a flower or something unexpectedly, I wish I could have another chance. When I can smell, [the sense] passes in a micromoment, and it is only once in a blue moon," Theresa says.[1]

THE FACTS ABOUT ANOSMIA

Smell disorders come in several forms. Anosmia is the absence of the ability to smell. Hyposmia is a decreased sense of smell. Dysosmia is an altered per-

Smelling in Stereo

How do we determine the location of a scent? To find out, one group of researchers led by Leslie Vosshall at Rockefeller University studied fruit-fly larvae that had scent-detecting neurons only on the left side of the head, only on the right side of the head, or on both sides. The investigators reasoned that, if two eyes and two ears help locate sights and sounds in three-dimensional space, two "noses" might do the same.

To test the larvae, the scientists used infrared light to visualize odor gradients, so they could see and measure where odor concentrations were the greatest. They found that a larva needs only one functioning olfactory neuron to turn toward a pleasant odor or away from a noxious one. But two neurons increase the "signal-to-noise ratio" of sensory input. Larvae with both left- and right-side smell cells find their way to a scented target more accurately than do their single-sided peers. Flies, Vosshall says, smell in stereo.[2]

ception of smell and includes the perception of an odor when no odor is present (phantosmia). It also includes an erroneous perception of an odor (parosmia or troposmia).

Smell disorders may be more common than most people—including physicians—realize. A Swedish research team tested nearly 1,400 healthy adults. They found some smell dysfunction in 19 percent of them (13 percent hyposmia and 6 percent anosmia).[3] The incidence of smell disorders is highest among older adults; it increases with age. A five-year study of nearly 2,500 residents of Beaver Dam, Wisconsin, ages fifty-three to ninety-seven, found some loss of the smell sense in one quarter of them. The prevalence increased with age, to the point where more than 60 percent of those of ages eighty and older were impaired. Men were more likely than women to experience smell-sense loss, as were smokers and the victims of stroke, epilepsy, chronic nasal congestion, or upper respiratory tract infection. Less than 10 percent of those affected were aware of their sensory loss.[4]

The source of anosmia varies among individuals. The problem can lie anywhere along the pathway from the nasal cavity to the recognition centers high in the forebrain. In some cases, allergies or chronic inflammation of the nasal lining or the sinuses interfere with airflow and keep odorants from entering the nasal cavity. Head trauma, multiple sclerosis, or Alzheimer's disease may alter the brain's perception of odorants. One study classified the cause of smell dysfunctions as 21 percent sinus and nose disorders, 36 percent occurring after infections of the upper respiratory tract, 17 percent after head trauma, and the remainder of unknown or miscellaneous causes.[5] Nasal polyps and diabetes are risk factors for anosmia, as is amnesia after head trauma that lasts longer than five minutes.[6,7]

Brain tumors are rarely the cause when smell loss is the only symptom. In cases such as Theresa's, in which head trauma produced the loss, MRI scans show injuries to the olfactory bulb and tract in 88 percent of cases, to the subfrontal region of the brain in 60 percent, and to the temporal lobe in 32 percent. Generally, the volume of the olfactory bulb is smaller in anosmic patients who experienced head injuries. Some doctors think the reason is damage to the axons as they run from the sensory neurons of the nose into the glomeruli.[8] Scent-detection abilities appear to relate directly to the size of the olfactory bulb. German researchers captured MRIs of the brains of thirteen patients who experienced partial loss of the smell sense. In the patients who improved over a period of nineteen months, the olfactory bulb of the brain increased in volume.[9]

How effectively a dysfunction of the smell sense can be treated depends a lot on its cause. Sinus surgery can remove blockages, but the sense of smell doesn't necessarily improve as a result. Inhaled steroids improve the smell sense in some patients, and systemic steroids help even larger numbers. Phantosmia can be treated, too. Blocking a nostril can sometimes relieve the perception of a continuous, unpleasant odor. Topical anesthetics and tranquilizers may provide temporary relief, and surgery can help in some cases.

REVERSING EVERYDAY ANOSMIA

Theresa's anosmia is irreversible because the scent-sensing region of her brain was permanently damaged. But the rest of us endure the loss of odor-detecting neurons every day and we recover nicely. How? Strong smells, colds, allergies, injuries—many things can damage sensory neurons inside the nose. They die like any other cell, but the lining of the nose has a remarkable ability to renew itself.

We have a large and friendly flock of stem cells to thank for our frequent and unconscious recovery from everyday anosmia. Many different kinds of stem cells get into the act, some of them replacing the neurons, others replacing structural components of the nose's lining. Researchers at Johns Hopkins University used fate-mapping techniques to find out where new smell-sensing neurons in the nose originate. (The idea behind fate mapping is that you "mark" a cell—with a dye, a radioactive tracer, or some other signal—and then see where the marker ends up.)

In mice, the researchers found that cells called horizontal basal cells (HBCs) lie idle most of the time—so idle that no one really expected them to be worth much. But the researchers found otherwise. While other kinds of stem cells handle "daily maintenance" and "minor repairs," HBCs swing into action when

Sniffing: Your Smell Machine

It goes something like this. You're strolling through a nighttime flower market in Bangkok when you catch an evanescent whiff of something alluring. *What's that?* you wonder, and sniff hard to learn more. You sniff repeatedly—moving purposefully now—and track that delicious fragrance to a stall bursting with *Plumeria*.

The sniffing was key; it set you on your mission. Without actively drawing air into your nose, you might have scarcely noticed the fleeting impression of a sweet floral scent. With it, you received a strong smell message and followed your nose to its source.

Scientists at the University of Pennsylvania School of Medicine wanted to find out why our sense of smell works better when we sniff. For answers, they turned to our old friend, the laboratory mouse, which, like a tourist in a flower market, tends to sniff, sniff, sniff when some interesting odor wafts through the air.

The researchers titillated tiny mouse noses with the scents of almonds and bananas in some experiments. In others, the mice got only the pressure of an unscented solution up their nostrils. The investigators found that the pressure was not quite as good as an odorant at stimulating scent-sensing neurons, but it definitely provoked a nerve impulse. The greater the pressure (within reason, of course), the greater the response. Furthermore, they found that odorant molecules and mechanical pressure work the same way; an intermediate messenger molecule called cAMP (cyclic adenosine monophosphate) activates the nerve impulse, whether the stimulus is scent or pressure.[10]

"The mechanosensitivity may increase the sensitivity of our nose, especially when stimulated by weak odors," says Minghong Ma, leader of the study and a professor of neuroscience at Penn. "We still don't know how it happens, but sniffing is essential for odor perception."[11]

major nose reconstruction is called for. They are the emergency squad that dashes to the rescue, regenerating both neuronal and nonneuronal cells after a major disaster.[12]

Without them (if we are like mice, anyway), we'd all be anosmic like Theresa. Perhaps, also, we'd value our sense of smell more. Theresa is one of the fortunate ones; she has learned to cope with her anosmia. Some people aren't so lucky. Anosmia can have serious psychological effects, including feelings of physical and social vulnerability and victimization. People with anosmia can develop eating disorders because their food loses its flavor. Anosmics can feel alone and misunderstood, with all the zest gone from their lives—both literally and figuratively.

Theresa hopes that her story will lead all of us to appreciate our sense of smell as she would—if only she could get hers back.

Chapter 9

The Sweaty Scent of Sex

Your doctor eyes you gravely. "I don't think you have a VNO," she says. "You do, however, have TAARs."

Is it time to check your life insurance policy and prepay the crematorium? No.

Your doctor's diagnosis tells you that you are a healthy human capable (probably) of responding to the chemical signals other humans emit. You may not have the VNO (vomeronasal organ, a structure in the nose) that other mammals use for that purpose. Or, if you do—and there is an area in your nose that is a likely candidate—it doesn't work the same way as it does in other mammals.

You do, however, possess an important class of receptors in your olfactory cells called TAARs (trace amine-associated receptors).[1] TAARs may well be the reason that women who live together menstruate at the same time. They may also explain why human males and females choose romantic partners with immune systems different—but not *too* different!—from their own.

All this has to do with pheromones, the chemical substances released by one individual that have some effect on the physiology or behavior of another. You don't smell pheromones the way you smell baking bread or smoldering rubber, but you respond to them anyway—unconsciously. Or do you? Some experts say humans have evolved past pheromonal effects. Others say pheromones are the invisible guests of honor at most of the 2.4 million weddings that take place each year in the United States, not to mention millions more worldwide.[2]

Let's look at the evidence.

NO VNO? NO PROBLEM

For a long time, the debate about humans and pheromones languished in the VNO pit. It is well known that other mammals respond to pheromones via the VNO—the impulses it sends acting as a direct line to brain regions that trigger significant behaviors such as mating. Anatomists and physiologists spent a lot of time arguing about whether humans have a VNO and whether it is functional. So, despite mounting evidence that chemical substances in the sweat and urine of one human affect the behavior of another human, the argument stalled. You can't respond to a chemical signal if you have no cells, tissues, or organs to detect it.

The VNO debate remains unresolved still, although a region of the human nose has been identified as the site for a VNO—but no one is sure whether it functions. Nevertheless, the argument moved away from the VNO in 2006 when Stephen Liberles and Linda Buck reported on TAARs. TAARs are receptors in olfactory cells, but you can't use them to smell the roses. No one knows yet what TAARs do in humans, but when Liberles and Buck studied mice, they found some specific pheromonal effects. One type of TAAR binds to a male hormone that prompts female mice to enter puberty. Several other types bind to sex hormones that are naturally excreted in the urine of males, but not females. Using TAARs, Liberles and Buck said, "[M]ice could, in principle, determine the gender and sexual status of other mice."[3] If TAARs can do this in mice, they might do something similar in people. Mice have fifteen TAARs; humans have six.

PHEROMONES AND THE MATING DANCE

The debate about the VNO and TAARs will probably remain unresolved for a while, but there's no shortage of evidence that something chemical is going on when first comes love, then comes marriage, then comes an older and wiser couple with a baby carriage, diaper bag, and second mortgage. Ladies, your higher brain centers may have a sense of humor, intelligence, and good looks in mind when you choose a partner, but your pheromones are looking for a good mate—one who will help you make strong, healthy babies. That's where the apparent ability of human females to assess the immune systems of potential sexual partners comes in. The key lies in a large segment of DNA called the major histocompatibility complex (MHC). The MHC codes for proteins that lie on the surfaces of cells. Those proteins allow the immune system to distinguish "self" from "nonself"—that is to tell one's own proteins (*Harmless. Leave them alone*) from the proteins of an invading virus or bacterium (*Danger! Attack and destroy!*). A major histocompatibility system in humans is called human leukocyte antigen (HLA). The more alike the HLA types of two individuals, the more alike are their

Scent and the Single Sperm

The same odors that attract male to female may also attract sperm to egg. In 2004 Leslie Vosshall of Rockefeller University reported that one odorant-receptor, called hOR 17-4, directs movement toward a mate. Vosshall tested tissue-cultured cells with hOR 17-4 receptors to see what odorants bound to them. She found a strong bond to bourgeonal, an odorant that the human nose–brain combo perceives as a pleasant floral scent similar to lily of the valley. Human sperm cells detect bourgeonal, too, although the brain doesn't know it. Watch human sperm under a microscope and see what they do when you add a drop of bourgeonal. They swim toward it. Human eggs probably don't produce bourgeonal, but they make something chemically similar—so similar that the molecule fits the hOR 17-4 receptor on sperm. Thus, as Vosshall puts it, "[M]ammalian sperm are indeed capable of 'smelling' their way to the egg."[4] (This begs further speculation: Could bourgeonal be used as a fertility enhancer, urging lazy sperm to get a move on? Could a bourgeonal blocker act as an effective contraceptive?)

immune systems; that's why HLA-typing is done to find a good donor–recipient match for transplanted organs.

Apparently, the female nose and brain can do a little HLA-typing of their own. University of Chicago researchers worked with forty-nine unmarried women, average age twenty-five. The women weren't using contraceptives and had never been pregnant. The scientists had six male volunteers wear T-shirts for two consecutive nights. The men avoided odorous foods, pets, cigarette smoke, colognes, and other strong scents. For the test sessions, the researchers put pieces of the T-shirts in foil-lined cardboard boxes. The women could smell the shirts through small holes in the boxes, but they didn't know they were smelling male bodies. To keep everybody honest, some of the boxes contained clean shirts (no scent) or shirts scented with chlorine bleach or clove oil.

The women selected the odor they'd most like to live with and the one they would choose to avoid. The researchers compared each woman's odor choices to her own HLA type. The results were dramatic. There was no preferred or disliked odor across the average of all women. Instead, odor preferences were unique to individuals and related to HLA type. Individual females selected as their favorite odors those that had some matches to their own HLA type, but not *too* many. The women rejected HLA types too similar or too dissimilar to their own. This detection was so powerful that women could discriminate the body odors of men who had but a single HLA gene different from their own.[5]

Men Who Stink

"He just came in from the gym smelling like a skunk, and he's forgotten to flush the toilet again. What a turnoff!" women complain, but not quite in unison. Although women generally have more sensitive noses than men, some females find the odor of male urine rather pleasant, and some don't notice that locker-room stench at all. The reason lies in the genes.

The odor receptor OR7D4 attaches to a molecule of androstenone, a breakdown product of the male hormone testosterone that's present in urine and sweat. There are two forms of the gene for that receptor. Call one *R* and the other *W.* Ladies, if you received the *RR* gene combination from your parents, chances are that an unflushed toilet offends your senses mightily. You may not be so offended, however, if you received the *RW* pair. Some women with this gene combo dislike the odor, but it doesn't drive them nuts. Others can't smell androstenone at all. Still others find the smell of androstenone quite pleasant; it smells like vanilla to them.[6]

As for that disgusting locker room scent, it all lies in receptor OR11H7P, which binds to a compound in sweat called isovaleric acid. If you have at least one functional gene for that receptor, you'll throw your sweaty sweetheart a towel and a bar of soap as he walks in the door. If you possess a mutated gene pair, you won't mind that locker-room stench a bit.[7]

Although some studies have produced slightly different results, most find some kind of link between scent detection in females and HLA type. And it does seem to be a uniquely feminine capacity. When Brazilian researchers had both males and females do the sniffing of opposite-sex T-shirts, the women performed as expected. The men showed no scent preferences for HLA types.[8]

A MOLECULE IN SWEAT

Much of the early research on pheromone effects in humans was done with sweat. Sweat contains many components, and it's impossible to know which one accounts for an observed response. Sweat also has a detectable odor, so it's hard to know whether the response is unconsciously pheromonal or consciously olfactory.

To simplify their experimental designs and fine-tune their interpretations, scientists have recently zeroed in on androstadienone (AND). AND is a steroid found in men's plasma, sweat, semen, and saliva. (Women have it, too, but in smaller amounts.) Several studies have found that AND reduces nervousness, tension, and other negative mood states in females.[9] It's not a conscious response to smell; in such experiments, all solutions are typically scented with a masking

"But I'm Gay . . ."

Similar findings apply. Gay people respond to pheromones, too, but with a different pattern of brain activity. Researchers in Sweden used PET to study brain responses to pheromones in homosexual men, compared to heterosexual men and women. The team used male AND and another compound EST, an estrogen derivative found in women's urine.

EST increased activity in the smell-processing centers in women, but not in the hypothalamus. In heterosexual men, it triggered increased activity in the hypothalamus.

AND showed the opposite pattern. It revved up the hypothalamus in women and the smell-processing regions in men.

In the Swedish study, the gay men's response to AND and EST was the same as the women's.[10] (Lesbian women have also been studied, but the results seem less clear.)

This suggests that the hypothalamic response to pheromones relates not to anatomical gender, but to sexual orientation. The question of cause and effect, however, cannot be resolved. Whether sexual orientation evokes the brain response—or whether the brain response shapes sexual orientation—remains a mystery.

odor—such as cloves—so the women don't perceive any particular "eau de man." Nonetheless, women who sniff AND report feeling more stimulated and in a better mood than those who sniff a control solution.[11]

Scientists at the Helen Wills Neuroscience Institute at Berkeley measured levels of the hormone cortisol in the saliva of women who smelled pure AND. The idea of this experiment was to see if AND could affect more than mood. Could it affect body chemistry as well? If did. Smelling AND produced elevated levels of cortisol in the women.[12] The level rose within fifteen minutes and stayed high for an hour. Cortisol is an arousal hormone; it's also associated with stress. (*Hmm. There's a paradox to ponder.*)

The study on cortisol doesn't settle the question of which comes first, mood or a chemical shift. The researchers report, "We cannot unequivocally determine whether AND influenced cortisol, which then influenced mood, or in turn, whether AND influenced mood through some nonhormonal mechanism, and the change in mood then led to a change in cortisol."[13]

It's possible that pheromones may not influence behavior directly, like a switch, but rather indirectly, like mood lighting. Perhaps pheromones don't trigger specific behaviors as much as they induce mood states that make particular behaviors more likely. This view gains support from research that uncovers only "context-dependent" effects of AND. In some studies, women show certain responses to AND only when the experimenter is male or a male is present in the room.[14] In other studies, effects are seen only when the test situation is sexually arousing, as in watching sexually explicit films.[15]

PHEROMONES AND THE BRAIN

When investigators look at brain activity, they find some AND-induced responses. One team of researchers used positron emission tomography (PET) scans to examine women's brain activity while the women performed a visual task. When exposed to AND during the task, the women showed heightened activities in brain regions associated with vision, emotion, and attention.[16]

The brain doesn't handle pheromonal cues the same way it handles other impulses that come in from the nose. For one thing, the processing is faster.[17] For another, the processing takes a different path through the brain. Researchers at McGill University in Montreal used PET to study the brains of individuals responding to ordinary odors as compared with the body odors of friends and strangers. The team found that a separate network processes body odors. Smelling a friend's body odor activated brain regions that handle familiar stimuli. Smelling a stranger activated the amygdala and insular regions, much the same as the patterns seen in studies of fear.[18]

The PET studies also reveal differences in how men and women respond to AND.[19] In women, brain activity increases in the hypothalamus region, not in the brain's usual smell-processing centers. The hypothalamus, which lies at the base of the brain, plays a big role in hormonal changes and sexual behavior. Men demonstrate the opposite pattern. They show greater activity in the smell-processing regions of the brain, and only there. This came as a surprise to researchers; they seldom see gender differences in olfaction.

TAARS AND THE VNO

Any way you cut it, the scent-of-sex discussion still pivots around the question of how humans can detect molecules that are not smelled but are somehow perceived. More evidence that TAARs may be the mechanism came in 2008 from an expected source: bird researchers in Germany. Like humans, birds lack a VNO. So the researchers combed libraries of chicken genes. The scientists found three TAARs of the same type as the human/mouse genes for TAAR1, TAAR2, and TAAR5. TAAR5 is known to be activated by compounds found in bird feces. So it's possible that TAARs are working in birds, humans, and other animals that lack a functioning VNO.[20]

Chapter 10

Express Train to Happiness

Aromatherapy speaks to something in my soul. I got hooked on it when I lived in England. I was teaching biology to more than one hundred highly hormonal adolescents daily, and my stress quotient was off the charts. A friend referred me to the local reflexologist and aromatherapist Mrs. Ophelia Bone. (I did not make up that name. There is an explanation. She wasn't born a Bone, so her parents are not to blame; she married a Bone, so her husband is.)

Mrs. Bone (I never called her anything else) had magic fingers and a magic cabinet full of sweet-smelling oils. She'd rub my feet, massage my aching muscles, and send me home with my personal mixture of oils, the concoction guaranteed to soothe away the miseries my students inflicted on me: lost homework, vandalized textbooks, failed exams, cheating (some), cursing (more), acid spills (one), and the complete range of crises that constitute the alpha to omega of a high school teacher's workday.

The potions worked, or so I believed and continue to believe to this day. Thus, because of Ophelia Bone's prowess, I am a believer in aromatherapy. I have no doubt that lavender soothes frazzled nerves, sandalwood relieves depression, peppermint improves memory, and lemon boosts sagging spirits. I have a cupboard full of essential oils to prove my devotion.

How might Mrs. Bone and I have fared had our devotion been tested in the scientific laboratory? Maybe not as well as we both believe.

PUTTING AROMATHERAPY TO THE TEST

It's one thing to believe in aromatherapy and quite another to demonstrate and measure its psychological or physical effects. Researchers have attempted to do both with mixed results.

For example, some evidence suggests that aromas can improve performance on various tasks. In one study, French researchers found that reaction times on simple tasks (responses to sight or sound cues) were shorter in an odor-infused environment compared to an odor-free condition.[1] The improvement occurred whether the odor was pleasant or unpleasant, perhaps because odors increase arousal levels, making subjects more alert.

Japanese researchers studied students who worked at a computer in one-hour intervals. During break times, the students smelled lavender, jasmine, or no odor at all. Like most of us, the students' concentration was poorest in the afternoon, when drowsiness sets in; but those who smelled lavender during their break performed better at that time. Jasmine and water had no such effect.[2]

Odors may influence how we view others. Scientists in the United Kingdom showed men's pictures to women, asking them to rate each man's attractiveness on a nine-point scale. While viewing the pictures, the women had pleasant (geranium or a male fragrance), unpleasant (rubber or body odor), or no odor in the room. The women rated the faces as significantly less attractive when they smelled the unpleasant odor than when they smelled either no odor or a pleasant odor. It didn't matter whether the odors were relevant to male bodies (the body odor and the male fragrance) or not (the rubber and geranium odors).[3]

Some evidence suggests that odors may even influence dreams. German researchers piped the odors of rotten eggs or roses into rooms where research volunteers slept. Subjects who slept in the foul-smelling place reported negative dreams. Those who slept with roses dreamed happy dreams.[4]

Emotional changes in response to odors can be tracked in the brain. French researchers used PET to monitor regional changes in brain activity when various kinds of cues were presented (compared with no stimuli at all). Whether the cue was sight, sound, or scent, pleasant and unpleasant stimuli—judged "emotionally charged" by the researchers—triggered increased activity in the brain's orbitofrontal cortex, temporal lobe, and superior frontal gyrus. Emotionally charged odors and sights (but not sounds) yielded greater activity in the hypothalamus and the subcallosal gyrus. Scent produced a unique result that sights and sounds did not: It increased brain activity in the amygdala, one of the brain's emotion-producing centers.[5]

Brain studies using functional magnetic resonance imaging (fMRI) have shown that the left side of the brain is more active in odor processing among right-handed people, and vice versa. Making a decision about odor pleasantness

Unhappy? Can't Smell a Thing

If smell can affect mood, can mood affect the sense of smell? Apparently yes.

A research team in Germany showed pleasant and unpleasant pictures to volunteers before rating performance on smell tests. After viewing unpleasant pictures, subjects rated test odors as less pleasant and more intense. After viewing pleasant pictures, odors too seemed more pleasant. The effects were related to mood changes, the researchers said.[6]

increases activity in the orbitofrontal cortex. During odor perception, the orbitofrontal cortex is more active in women than men, which may explain, in part, why women are usually better at identifying odors than are men. Unpleasant odors activate the amygdala and insula more than pleasant odors do, perhaps because unpleasant odors are more emotionally arousing.[7]

AROMAS AND PAIN

Things get more complicated when we get to the subject of pain. It's possible that scent can diminish pain—in fact, some researchers find that it does. The reasons why are unclear. Perhaps odors distract attention away from pain. Researchers at the McGill Centre for Research on Pain in Montreal used heat-induced pain and pleasant and unpleasant odors to manipulate the attention and emotional state of volunteers. Shifts in attention between heat and smells did not alter mood or anxiety, but when people paid attention to their pain, they judged it as more intense than when they paid attention to the odor.[8]

Or maybe a scent puts the pain sufferer in a better mood, so the pain seems less distressing. In one experiment, researchers piped the scent of orange or lavender into the waiting room of a dental office. They measured the effects of the scents on anxiety, mood, alertness, and calmness compared to music alone and a control environment (no odor, no music). They found that both the orange and the lavender reduced anxiety and improved mood compared to music alone or the no-music–no-scent condition. The researchers concluded that fragrance can alter emotional states and make dentistry more tolerable.[9]

A study at the University of Florida College of Dentistry compared how subjects rated heat and pressure pain after inhaling lavender, rosemary, and distilled water (the control). Subjects didn't rate their pain any differently during the session, but after the session they were more positive. They remembered their pain as less intense and less unpleasant when they had smelled the lavender—and, to a lesser extent, the rosemary.[10]

Odors and Memories

Talk about instant travel in space and time! It seems nothing can take you back to a memory faster than an odor. The experience is called the Proust phenomenon. It's named for the French novelist who described a flood of childhood memories that came back to him after he smelled a madeleine pastry dipped in linden tea.

It's hard to argue with Proust's idea. In our everyday experience, it seems that nothing jogs the memory better than a smell. To qualify as the Proust phenomenon, however, an odor-prompted memory needs to be sudden, autobiographical, old, vivid, and emotional. There's evidence for all those things. In a series of experiments, Swedish researchers presented nearly one hundred older adults with one of three cue types: a word, a picture, or an odor. They asked each person to describe any autobiographical memory that came to mind after the cue. The scientists found that odor cues triggered older memories than the words and pictures did. Most of the odor-cued memories came from the earliest years of life, when the people were younger than ten. Memories evoked by words and pictures came mostly from adolescence and young adulthood, ages eleven to twenty. Odor-prompted memories carried with them "stronger feelings of being brought back in time and had been thought of less often than memories evoked by verbal and visual information."[11]

In another series of experiments, the Swedish researchers gave seventy-two older adults one of three cues: odor only, name only, or odor-plus-name. Again, they asked their subjects to describe autobiographical memories. As before, the smell-induced memories carried with them the most intense emotional responses, but words got in the way. The odor-plus-name condition dampened the trend toward memories from the earliest years. Subjective ratings of pleasantness and feelings of going back in time decreased when odors and their names were presented together.[12]

The big question, however, is whether odor is better than sight or sound at bringing back memories. Rachel Herz, a psychologist at Brown University, says it isn't. She used words, pictures, and odors of campfires, fresh-cut grass, and popcorn to elicit autobiographical memories in volunteers. She found that odor-evoked memories were more emotional and evocative than those prompted by the other cues, but she found no difference in the vividness or specificity of the memories:[13]

> What I found in these experiments is that, in terms of their accuracy, detail, and vividness, our recollections triggered by scents are

> just as good as our memories elicited by seeing, hearing, or touching an item—but no more so. Yet our memories triggered by odors are distinctive in one important way: their emotionality. We list more emotions, rate our emotions as having greater intensity, report our memories as being more emotionally laden, and state that we feel more strongly a sense of being back in the original time and place when a scent elicits the past than when the same event is triggered in any other way.[14]

The structures and functions of the brain explain the Proust phenomenon, Herz thinks. She and her team took fMRI pictures of women's brains while the volunteers retrieved personal memories. The scans showed greater activity in the amygdala and hippocampus when memories were triggered by an odor—as compared to other cues to memory. The women's emotional responses were greatest with odor cues, too.[15] Both the amygdala and the hippocampus play important roles in memory formation. Both are centers for emotional processing as well.

However, not all studies confirm that odor relieves pain. Smells may, in fact, make pain worse. In one study in the U.K., volunteers dunked their hands and forearms in freezing water while smelling a pleasant odor (lemon), an unpleasant odor (machine oil), or no odor. They rated the severity of their pain at five-minute intervals for fifteen minutes. At five minutes, those smelling either a pleasant or an unpleasant odor reported greater pain than did the no-odor subjects. At fifteen minutes, pain was greater in the unpleasant-odor group. Thus, in this case, odor didn't relieve pain; it exacerbated it.[16]

Some studies suggest that the pain-relieving effects of odor—if they exist—vary by gender. Canadian researchers reported that pleasing odors produced a positive mood in both men and women. Unpleasant odors produced a negative mood in both sexes. But pain relief was different. Only women gained pain relief from pleasant scents.[17]

AROMATHERAPY REVISITED

The team of researchers at Ohio State University who recently took an objective look into aromatherapy might have accepted Mrs. Bone and me into their study. Some of the fifty-six male and female healthy volunteers for the aromatherapy research project were believers. Others were skeptics, or they just didn't have much of an opinion about aromatherapy one way or another.

The Scent of Impatience

Walk into many supermarkets and you'll immediately notice the scent of freshly baked bread or pastries. In some cases, the fragrance is not coming from the bakery. It's actually piped into the ventilation system and blown at customers as they enter the door. There's a reason. Good smells do more than make you hungry. They make you eager to get what you want *now* . . . and the object of your desire needn't be food!

As part of her doctoral research at the University of Toronto, Singaporean Xiuping Li asked thirty-six females how happy they would be if they won $100 in the lottery now, in three months, in six months, or one year to two years from now. The women rated their happiness of a scale of 0 to 100. Half of the participants answered the question in an unscented room. Half worked in a chocolate-chip-cookie–scented environment.

The cookie-scented participants devalued delayed gains. They wanted their money immediately and were less happy than those in the unscented environment about future wealth. When tested for the likelihood of making an unplanned and unnecessary purchase (such as a nice sweater), 67 percent in the cookie-scented room were more likely to do so—compared to only 17 percent in the unscented room.[18]

The subjects wore cotton balls infused with lemon oil, lavender oil, or distilled water taped below their noses during three days of testing. Heart rates and blood pressures were monitored. Blood samples were taken. Subjects took psychological tests of mood and perceived stress; they played word games that evaluated their emotional states. Their ability to heal was measured on a patch of skin to which tape was repeatedly applied and removed. Reactions to pain were gauged when the subjects put their feet in ice water.

It sounds like a tough regimen to me, but I'd have made it through, and I think Mrs. Bone would have, too—for the sake of science. Neither of us would have been too keen on the results, however. Aromatherapy failed most of its tests. Neither lemon nor lavender had any effect on pain ratings, stress hormones, wound healing, or the immune system. Sadly, in some cases, distilled water performed better.

One ray of light emerged from the darkness, however. Lemon improved the self-reported mood of the volunteers. (After all they went through, I think *something* should have!) The heightened happiness occurred regardless of expectations or subjects' previous experiences and their views on aromatherapy.[19] Mrs. Bone and I would have cheered loudly at that.

Name That Scent

Calling a rose by another name *does* change its smell. That's the conclusion of researchers at the Montreal Neurological Institute. They had forty people smell fifteen different odors on three separate occasions. The odors were inherently pleasant, unpleasant, or neutral, but they were presented with different names. Positive names were labels like banana bread, peanut butter, or spearmint gum. Negative names included dry vomit, old turpentine, or insect repellant. Neutral names were simply two-digit numbers such as 53 or 28.

The researchers measured their subjects' skin conductance, a generally reliable measure of an emotional response. Subjects also related how pleasant and intense they thought the odors were. Names affected all the outcomes, but most strongly the judgment of pleasantness. The same odors were judged more pleasant when presented with a positive name than when presented with a neutral or negative name. Intensity ratings were affected, too. The people rated odorants as stronger when they were linked with negative names than with positive or neutral words.[20]

Physical measures bore out the differences in ratings. Skin conductance increased when odors were presented with positive or negative names. The scientists concluded that a rose by the name "rotting flower" would not smell as sweet.

So I won't clean out my aromatherapy cupboard quite yet, and I doubt Mrs. Bone will either. "The human body is infinitely complex," says William Malarkey, one of the authors of the study. "If an individual patient uses these oils and feels better, there's no way we can prove it doesn't improve that person's health."[21]

PART THREE

TASTE

Taste's importance to our daily lives is self-evident in its metaphors—for example the "sweetness" of welcoming a newborn child, the "bitterness" of defeat, the "souring" of a relationship, and describing a truly good human as the "salt" of the earth.

—Paul A. S. Breslin
Monell Chemical Senses Center
2008

Chapter 11

The Bitter Truth

I remember my seventh-grade science class. We were all seated around wet sinks and gas jets, armed with cotton swabs and little bottles of clear solutions labeled sweet, sour, salty, and bitter. Our mission: to map our tongues. By swabbing small amounts of the lemon juice and sugar water onto different areas of our tongues, we were supposed to "discover" the tongue map shown in our textbook. Our book's diagram showed which areas of the tongue respond uniquely to which flavors. If you couldn't match the illustration in the book, you weren't doing the lab right or there was something wrong with your tongue.

I was careful about how I dipped and swabbed, so there was definitely something wrong with my tongue. I mostly tasted nothing, tasted the same thing no matter where I put the swab on my tongue, tasted something different from what the bottle said, or managed to gag myself on the swab. Surreptitiously (because Miss Chedister tolerated no slackers), I glanced around the room. It looked as if the same thing was happening to everybody else, even Jenni Simpson who never got anything less than an A+ and Wayne McArthur who didn't know his tongue from his ear. We all did pretty well on that lab, though. With our colored pencils, we copied the tongue map from the textbook onto our lab reports. Chalk up another breakthrough in science.

I came away from that experience vaguely uncomfortable with science. Sure, the orderliness of the universe—and the sciences that describe it—remained unmarred. My tongue map looked like it was supposed to, and so did everybody else's. But it was all a tacitly agreed-upon lie—my tongue told me that—and if science isn't truth, it isn't anything—at least, that's what I thought in the seventh grade. Maybe I still do.

TASTING THE TRUTH

Some of us carry our seventh-grade obsessions into our dotage, so I never got over the bitter betrayal that stuck on my palate when I learned, a few decades later, that tongue maps are bunk. "There isn't any 'tongue map,'" says Yale's Linda Bartoshuk, and she should know.[1] She spent years asking people to suck on little pieces of paper that contained a harmless chemical called PROP (6-n-propylthiouracil). Some people can't taste PROP at all. Others find it slightly bitter. But some people are a threat to Bartoshuk's life and career; the taste of PROP is so bitter for them that they want to bite somebody, and she is nearby.

Bartoshuk swabbed blue food coloring on the tongues of her hapless victims (in the name of science, of course). Her blue food coloring stained areas of the tongue that lack taste buds. The fungiform papillae (where many of the taste buds are located) stay pink. (They are called fungiform because they are shaped like mushrooms.) She found that people who don't taste PROP have the smallest number of fungiform papillae. Those who taste it the most have the largest number.[2] It's the number that matters, not the location—and the number comes from your parents. Get too many taster genes from your mom and dad and you dislike PROP so much that you won't eat your spinach, cabbage, or Brussels sprouts. They taste too bitter.[5]

So who perpetrated this tongue-map hoax on three generations of seventh-graders? Blame the Germans . . . and Harvard. In 1901 a fellow named Hanig published his Ph.D. dissertation. In it, he drew a numberless graph that showed certain areas of the tongue to be more sensitive to certain tastes than others. Forty years later, Harvard professor Edwin Boring manipulated Hanig's data into a graph with real numbers that looked impressive. The tip of the tongue tasted sweets, his graph seemed to show. Bitter-taste sensitivity lay in the back. Sour tastes registered along the edges, while saltiness worked just about everywhere.

The map failed to acknowledge the fifth flavor, umami, which is generally accepted today as the reason why rib-eye steaks and baby back ribs are so popular (and why cholesterol-lowering drugs are so widely prescribed). The map survived because it was simple to use and pretty to look at, until a new generation of molecular scientists showed that all taste buds everywhere on the tongue and palate respond to all flavors. As for tongue maps, "Your brain doesn't care where taste is coming from in your mouth," Bartoshuk says.[6]

IT'S ALL IN THE RECEPTORS

Taste buds are clusters of about fifty to one hundred taste cells, and it's not the bud that's all that important, but the individual cells. Since you probably don't have access to PROP, get some of your taste cells going by putting a spoonful of dry instant coffee in your mouth. Unless you are ageusic, your mouth will

Defining Taste and Flavor

Scientists insist on a distinction between taste and flavor. Taste impulses begin in the taste buds and are usually said to be limited to five: sweet, salty, sour, bitter, and umami (or savory). Umami, which confuses everybody who isn't Japanese, has been described as "brothy" or "meaty." It is said to be the taste of amino acids or protein foods in general, or the amino acid glutamate in particular. The best way to appreciate umami is to sprinkle some Parmesan cheese on your tongue.

Flavor is more complex than taste alone. It results from a blending of taste with smell, as well as touch detectors that inform the brain of a food's temperature, crunchiness, chewiness, sliminess, or creaminess. That's why lemon meringue pie, lemonade, and lemon chicken taste hardly anything alike.

Fat, which everybody loves, has always been considered a flavor, not a taste—mostly dependent on smell or "mouth feel." But that view is disputed. A few years back, researchers at Purdue University found that the levels of fat in the blood rise when we do nothing more than put fat in our mouths—no swallowing or digesting required.[3] To learn more, the scientists recruited some volunteers who took safflower oil capsules after fasting overnight. After that, some tasted cream cheese but didn't smell it; some smelled cream cheese but didn't taste it; some both smelled and tasted it; some did neither. Those who both tasted and smelled the cream cheese had blood fat levels triple that of the controls. But fat levels rose as much in those who only tasted (didn't smell) the cream cheese as in those who did both. Blood fat didn't rise above the baseline in people who smelled the cheese but didn't taste it. The conclusion? Fat may someday join the "big five" in our basic menu of tastes.

Another taste sensation is jockeying for position seven on the list. In 2008 researchers at the Monell Chemical Senses Center in Philadelphia announced that mice can taste calcium. They studied forty different strains of mice, some of which lap up calcium water like it's ice cream. The scientists found two receptors on the tongue that bind calcium; one is T1R3, a receptor better known for responding to sweet tastes. Because mice and humans share many of the same genes for taste receptors, the Monell researchers think humans may be able to taste calcium, too.[4]

pucker, and you'll have no trouble recognizing the coffee's bitter flavor. That happens because of a "family" of protein receptors in your taste cells inelegantly named T2R1, T2R2, T2R3, and so on. What's a receptor? It's like a slip for a boat. It's a little docking place where chemicals can "tie up" for a while.

The T2Rs are found (or not found) in various taste cells. The T2R receptors are tailor-made for latching onto the molecules of all things bitter. When submicroscopic bits of bitterness tie up to T2R docks, taste cells start chattering. Only a few of the individual cells have a direct connection to a nerve, but the entire taste bud does, so all this molecular "talk" forces the taste bud to send a signal. With all that dry coffee in your mouth, the brain responds with the neurological equivalent of a grimace. Since most things bitter are also poisonous, you'd best give up coffee. (Or maybe not. There are exceptions to every rule.)

Bitter isn't the only taste that works that way. Each of the five basic flavors—bitter, sweet, sour, salty, and umami—has its own specialized receptors and its own specialized cells where specific receptors operate, say biologists at the University of California at San Diego. UCSD scientist Charles Zuker (now at Columbia University Medical Center) and his collaborator Nicholas Ryba of the National Institutes of Health have identified about thirty receptors for bitter, but they have found only three for sweet and umami. They named them T1R1, T1R2, and T1R3. The T1Rs work together, not independently. The main reason we taste sweets is that T1R2 and T1R3 reside in the same taste cells and play together well. High concentrations of sugars, but not other sweet things, are handled by T1R3 alone. The umami taste receptor appears to be a complex of two receptors, T1R1 and T1R3. The complex is shaped like a Venus flytrap. When the amino acid glutamate, which is believed to be the source of the umami taste, lands inside the flytrap, the flytrap closes around it, and the umami taste sensation is born.[7] Zuker, Ryba, and their colleagues have found the receptor for sour, too; it is called PKD2L1. They haven't found the salty receptors yet, but they are working on it.

It's all as neat and tidy as the tongue map in my seventh-grade textbook—and equally unsatisfying. I've always been in love with science, but I've learned that even those we adore have their flaws. If scientists could accept for decades a misinterpretation of a graph that had no numbers, I'm suspicious of pat answers about boats at docks that explain everything from bitter English ale (which I love) to bitter rotten tomatoes (which I don't) in two paragraphs. I want to know, *How do they know that?* Can those researchers in San Diego and Bethesda see little bitterness molecules tying up like boats? Can they see impulses traveling along nerve cells like waves along a slinky toy? I don't think so.

What they can do, however, is identify which genes (pieces of DNA) code for which proteins. They can then breed mice that don't have those genes so the animals don't make those proteins. They call these mice "knockout" mice, because the researchers can remove (knock out) a particular gene and then see how it affects the mouse's health, appearance, behavior—or, in this case, taste preferences. The scientists can also give to the mice particular genes the animals wouldn't naturally have—in this case, human taste receptor genes.

Taste researchers measure taste preferences in mice by counting how many times the animals lick at some flavored water, as compared to plain water. More licks: the mice like the taste. Fewer licks: they are tasting something bad. So in one experiment, Zuker and Ryba knocked out the gene for the T2R5 receptor-protein that latches onto the compound cycloheximide. Both humans and mice normally find that substance disgustingly bitter. But mice that didn't have T2R5 in their taste cells didn't care if cycloheximide was in their water. They licked away happily, oblivious to the bitter taste.

In another experiment, the research team engineered mice to make T2Rs from humans that mice don't normally have. Those particular T2Rs are the reason PTC (phenylthiocarbamide) and salicin taste bitter to most humans. The engineered animals turned away from their PTC- and salicin-laced waters just as humans would—and ordinary mice wouldn't.[8]

So far so good, but that's only cause by association, isn't it? To prove that the T2Rs are the receptors for bitter substances, Zuker and Ryba had to prove that bitter compounds were docking in the T2R slips. They couldn't do that in living animals, but they were able to in cells grown in culture dishes. Their research confirmed that at least one category of bitter taste receptors had been identified, but there was a lot more to learn.[9]

FROM MOUTH TO BRAIN

Life would be a lot simpler for taste researchers if single taste cells contained only one kind of receptor and responded to only one taste. But things are never that simple. One complication is that the cells in taste buds that have receptors aren't the same ones as the cells that send taste messages to the brain. Therefore, chemical messengers must get the requisite information to the impulse-sending neurons and get a signal started. Two candidates for the messenger role are the neurotransmitters ATP (adenosine triphosphate) and serotonin. Only receptor cells release ATP, and only impulse-generating cells release serotonin. Those substances (among others probably) carry out the cell-to-cell signaling needed to get a taste message traveling.[10]

There is another complication. Taste cells can contain more than one kind of receptor and neighboring taste cells "talk" to one another. The brain's ability to sort out the message from all this chatter—between bitter and sweet, anyway (salty and sour are probably different)—may lie in another category of chemicals, some short chains of amino acids called peptides, which are the building blocks of proteins. Two molecules of interest are neuropeptide Y (NPY) and cholecystokinin (CCK). Both are present in the same taste cells.

A Taste for (Tasteless) Calories

The brain has a mind of its own when it comes to recognizing and discriminating tastes. A multistate team of brain scientists led by Walter Kaye at the University of California at San Diego did MRI brain scans of twelve healthy women as they consumed varying concentrations of sucrose (real sugar) and sucralose (artificial sweetener). The scientists found that both sucrose and sucralose activated the same primary nerve pathways that send "sweet messages" to the brain, but sucrose prompted a stronger response in the brain's anterior insula, frontal operculum, striatum, and anterior cingulate. Furthermore, only sucrose stimulated production of the neurotransmitter dopamine in midbrain areas seen to "light up" when the women judged the sweet taste as pleasant. Those results, the researchers say, show that the brain distinguishes high-calorie sweeteners from calorie-free substitutes, even if the mouth does not.[11]

A research team at Duke genetically engineered mice to lack the TRMP5 ion channel in taste receptor cells. Without it, the animals couldn't taste sweet, bitter, or umami. The mice showed no interest in artificially sweetened (no-cal) water, but they demonstrated a marked preference for sugar (hi-cal sucrose) solutions, a preference based solely on calorie content. Recorded signals from the animal's neurons showed that the high-calorie food produced a surge of dopamine release in the reward-seeking centers of the brain. The more the better: the higher the caloric content, the greater the surge. Fattening foods, the investigators concluded, have an express train ticket to the brain's motivational circuitry, even when their flavors can't be perceived.[12]

Scott Herness, a neuroscientist at Ohio State University, thinks NPY and CCK are important communicators of taste information. He says that NPY and CCK may initiate different signals traveling to the brain, depending on what substances are binding to receptors at the moment. NPY makes the taste bud send a sweet message. CCK triggers a signal for bitterness, Herness says.

The way Herness and his team figured this out is clever. They isolated single cells from the taste buds of rats, then attached microscopic electrodes to the taste cells. The electrodes let the researchers detect the cells' electrical activity. Then they applied either CCK or NPY to the cells. "NPY activated a completely different signal than CCK did, suggesting that the peptides trigger completely different responses in individual cells," Herness reported.

In another experiment, Herness stained taste cells and used fluorescent light to study them; the staining and the light showed which cells were making NPY and CCK. Only certain taste cells had the peptides, but without exception, if a cell had one, it had the other.[13] "That surprised us, too," Herness said. "It may be that these cells release both peptides when something sweet or bitter is on the tongue. CCK might excite the bitter taste and at the same time inhibit the sweet taste, so the bitter message gets to the brain."[14]

This suggests that the perception of taste or flavor (a combination of taste, smell, texture, heat, and more) is only partly in the mouth. It's the brain that actually does the tasting—no one disputes that—but how the brain makes up its mind is debatable. Whether the brain processes a large number of straight-line messages in a one-for-one fashion or responds to complex patterns of input from sensory neurons is the subject of hot debate among taste researchers.

"Somewhere up the neural processing pathway there is an integration of taste information," Zuker says. "We ultimately want to know how the brain integrates the various inputs from the taste system—bitter, sweet, umami, salt, and sour—and uses that information to trigger selective behavioral and physiological responses."[15]

MAPPING THE TERRITORY

This is where the brain mappers come in, at least in determining the path that signals follow. The electrical message from a taste receptor stimulates a response in a sensory neuron that's a mainline to the brainstem at the base of the skull. The fibers of these neurons run through several of the cranial nerves. The cranial nerves carry taste information to a region of the brainstem called the nucleus of the solitary tract. From there, secondary or relay neurons pass the taste message on to the thalamus (which acts as a sort of central dispatch station) and to two other brain regions: the limbic system and the cerebral cortex.

The limbic system is a general name for a number of different structures including the hippocampus, hypothalamus, and amygdala. In these brain parts, memories are formed and emotions are generated. Here's where you experience

Smell? Taste

Smell cells and receptors differ from similar structures in taste buds in two important ways. While taste cells may have functioning receptors for several different tastes, olfactory cells possess only one. Expression of genes is tightly controlled so that each smell cell is a specialist. Another difference is that taste-receptor cells don't generate a nerve impulse directly; they pass that job on to other neurons nearby. Smell cells are different. Each sends an axon directly to the olfactory bulb of the brain.[16]

a negative reaction to bitter and a positive reaction to sweet, and you'll learn to avoid or seek out certain foods or flavors next time, although you may not be aware of the reason for your attraction or revulsion.

In the cerebral cortex, brain activity increases in response to taste in the primary taste cortex, which includes regions called the anterior insula and the frontal operculum (at the side of the head, just above the temples). Brain scans also show activity in the cingulate cortex a bit higher up. These are the regions that tell you that your coffee is too bitter or too sweet. Here you can make the conscious decision that you want more sugar in your coffee—or, if you've learned to like bitter—you prefer it black.

Another region of the brain that gets into the act is the orbitofrontal cortex, which lies just behind and above the eyes. Here taste information feeds into the reward assessment and decision-making systems. Taste messages are connected in this area to the "temporarily organized behavior" of food- or sensation-seeking. This is a region where we pay attention and act toward a goal.[17] Positive and negative judgments about tastes are both made here, although in slightly different areas.[18]

My seventh-grade tongue map days are a distant memory, but my cerebral cortex now pines to retrace the route I've taken since then. I think I have come full circle. I've abandoned tongue maps in favor of another sort of map that follows a sensory path from tongue to brainstem to emotions, memory, and cognition. Both maps are oversimplifications. Receptors, nerve pathways, and brain processing centers are ways and means, but they don't explain the essence of the taste experience. I have a new kind of map, but it is only a map. It charts a territory, but it is not the territory itself. The bitter truth is, I know more than I did in the seventh grade, but I still don't know much.

Chapter 12
Coconut Crazy

My daughter Elise and I sit on the beach, slathering ourselves with pints of sunscreen because all the experts say tanning will kill us. Still, the sunshine feels warm and our sensibilities mellow. It's a golden mother–daughter moment. Elise, at twenty-five, is a rising corporate executive. Our relaxed times together are precious and few.

Abruptly, a dedicated sun worshiper in a yellow string bikini invades our space. With a flourish, she spreads her towel inches from ours and plops to the ground. Thumbing her nose at the melanoma statistics, she begins basting her body with a coconut concoction guaranteed to fry her milky skin to a burnt sienna in minutes. Its scent reminds me of warm, creamy pies and tangy piña coladas. Elise has a different response.

"Can we move?" Elise asks, looking a little green despite the rose tint of my sunglasses. She furrows her brow. Her lips curl.

"Sure," I reply, shoving my paperback novel into my tote bag and slipping on my flip-flops. Elise is up and running already, a good thirty yards down the beach before I catch up. "Are you all right?" I inquire with motherly concern.

"It's the coconut," she says, punctuating her reply with a gasp and a choke.

Once we've settled onto our towels again, she reveals all. Her coming-of-age trip was a spring break to Tenerife in the Canary Islands with two dozen classmates. One evening, as the tropical sun set and the revelers grew merry, the libation of choice was a bottle of Malibu . . . or maybe two. Malibu (it was news to me) is an extremely sweet, coconut-flavored liqueur—hardly the drink most people would choose for purposes of inebriation—but these were college kids and they procured what they could. The consequences the next morning were the predictable pounding head and churning stomach, but Elise was determined not to miss the scheduled boat ride around the island. The combination of high seas and Malibu morning was too much for her digestive system.

"Never have I retched like that, before or since," she admits.

I put my hand up to stop her. I don't need to hear the nauseating details. My daughter is a victim of what scientists call hedonic dysfunction. The taste—even the smell—of coconut repulses her. She exhibits aversive behavior. She runs. Her brain was permanently altered by overconsumption, and her conscious mind has extended her midbrain response far beyond the initial stimulus. While many people relish coconut in all its forms, the mere mention of the word turns her stomach.

I reassure her lamely, confessing that I have the same response to sweet potatoes, although for very different reasons. We don't bother to get into them, but I start to wonder. What happens in our brains when we love or hate a taste? Is this a simple conditioning response or do we—with too much Malibu or something else—permanently alter our synapses?

HEDONISM AND THE LAB RAT

To find out, I leave the beach and fly off to the University of Michigan in hopes of meeting Kent Berridge and Wayne Aldridge, two researchers who have done some exciting work on hedonic dysfunctions like Elise's. Berridge is a cognitive psychologist who runs an animal laboratory where various kinds of learning studies are conducted in rats. Microscopic electrodes implanted in the animals' brains record the firing of neurons when they taste something they like or something they don't. Berridge is out of town, but one of his graduate students, Steve Mahler, takes me on a tour of the lab.

We enter a large room, where a half-dozen students and technicians work at computer stations. Off the main room are smaller experiment rooms equipped with transparent and mirrored boxes. In them a rat's every whisker-twitch can be videorecorded and scrutinized. On one desk, a TV monitor runs a recorded sequence of rat behavior. Taped above the monitor is a message, "Have you scored today?" Students and research assistants must catalog many hours of recordings daily, carefully coding the time sequence of every movement a rat makes during an experiment.

Not all the rats are video images. Some live ones are in the lab today, being readied for various experiments. One young woman holds a rat to her chest, cradling it as she would a baby while she gently administers an injection. I like the rats. They are white and clean. They appear warm, well fed, and comfortable in their experimental environment, but I'm under no delusions. They will give up their lives so that we can acquire knowledge, and I approve their sacrifice. Some things cannot be learned from test tubes, microscope slides, or computer simulations. Those who condemn the use of animals in all scientific experiments wear blinders, in my view.

I pull myself back from my digression and ask Mahler to describe the work. "How can you tell if a rat likes or hates a taste?" I ask.

He tells me that rats make faces about foods much the same as people do. Like most humans, rats adore sweet tastes. When they drink sugar water, they signal their approval by protruding their tongues forward and sideways. Sometimes they lick their paws, too, as if hoping to lap up the last drop. Mahler calls these behaviors positive hedonic responses. In contrast, the neutral taste of plain water elicits a *comme ci, comme ça* reaction. The animals move their mouths rhythmically, and that's pretty much all. They don't love the water, but they don't mind it either.

Aversion is a different story and a very big deal. When rats hate a taste, they may shake their heads, flail their forelegs, wash their faces, or rub their chins—but most of all they gape. A gaping rat is a sight to see, and there's no better way to elicit a gape than give the animal a strong salt solution. Water that is too salty causes rats to gape nearly every time.

In their experiments, Mahler, Berridge, Aldridge, and their coworkers have used these distinctive behaviors to find out what's going on in rats' brains when the animals taste flavors they like and flavors they don't. Along with observing behaviors, the researchers study the firing of neurons in a region of the brain called the ventral pallidum (VP).[1]

Mahler pulls an oversized atlas of brain structure from a shelf and I get a short course in brain anatomy. The VP is part of the brain's basal ganglia, several masses of gray matter fondly nicknamed the brain's "basement." Nestled snugly between the bigger and more famous cerebral hemispheres, the ganglia conduct a lot of business, including controlling involuntary movements. Damaged or inoperative neurons in the ganglia cause disorders of movement such as Parkinson's disease, but that's not all that goes on there. The ganglia communicate with many other parts of the brain, including those centers that specialize in emotions and decision making.

EXPERIMENTING WITH TASTE

I'm eager to learn more, and I thank Mahler for the lab tour. Although I've missed Kent Berridge, I've had better luck getting an appointment with Wayne Aldridge. He meets me at a Starbucks near campus. (There's nothing coconut on the menu board, not even a macaroon to go with my lowfat latte.)

Aldridge and I settle at a table to chat. He's good company, a neurophysiologist who never lost his interest in behavior, he explains. He doesn't handle or observe the rats directly. Instead, he makes recordings of neuronal firings in the animals' brains and painstakingly identifies the signature signals from single neurons. Just as people can be recognized by their voices, individual nerve cells speak with a distinctive waveform that Aldridge can capture and interpret. It's the pattern of those signals that intrigues Aldridge. He's trying to crack the "neural code," the pattern of activation that, something like Morse code, uses single impulses in various combinations to create the brain's equivalent of letters, words, sentences, and paragraphs.

Aldridge reviews for me the experiment that brought me here. A few years back, he, Berridge, and several students started out by implanting electrodes in the VPs of eight male rats. The research team went to a lot of trouble to make sure that they put the electrodes in exactly the right places and that the sensors would detect and record the firings of precisely 167 individual neurons. The procedure didn't harm the rats or change their behavior in any way, Aldridge assures me, and neither did the tubes that fed the rats measured amounts of plain water, sugar water, and salt water at carefully timed intervals.

The next step was to make and analyze video recordings of behaviors after the rats drank the solutions. The purpose was to confirm that these rats weren't somehow different from all the other rats in the world. Sure enough, there wasn't a maverick in the bunch. The rats protruded their tongues (sugar), moved their mouths (water), and gaped (salt) right on cue.

Now comes the elegant part of the experiment, the part that reveals more than simple taste preferences. The Michigan team changed the rats' normal response to salt water by depleting the animals' bodies of salt. In a salt-deprived state, brought on by a combination of drugs and a salt-free diet, the animals started to like the salt solution. They stuck out their tongues and licked their paws, just as they continued to do for sugar water. Their once aversive response to salt flipped end over end. The unpleasant turned pleasurable.

Aldridge explains what he learned from examining the firing of neurons in the VP. The more the animals liked a solution, the faster the neurons in the VP fired. In the beginning, the response rate for salt was low—much lower than for the pleasurable sugar. But when the animals' bodies were salt-deficient, their VP firing rates doubled, equaling or exceeding the rate that sugar water evoked. What's more, the firings in the VP matched up with the behaviors. When happy rats protruded their tongues, the firing rate was high. Gaping rats (I imagine them shouting, "Yuk!" if they could) had low VP firing averages.

"We converted something that wasn't pleasing to something that suddenly became pleasurable, and when we did that the neurons we were studying switched their response," says Aldridge. The VP is a "hedonic hotspot," he adds. The faster its neurons fire, the more pleasure the animal experiences.

LIKING AND WANTING

Aldridge tells me that Berridge is famous for having worked out the distinction between liking and wanting. I can't see much difference, so Aldridge serves

Defect in Her Dopamine?

Aldridge is suitably cautious about jumping to conclusions, but based on what he's told me, I'm wondering if dopamine might not have something to do with Elise's Malibu aversion. True, neurotransmitters have different effects depending on where they operate in the brain, but dopamine is also known to be a big player in the drama of conditioned learning. Neurons that use dopamine start kicking out the stuff when a reward is offered at the same time as a stimulus; that's the classic conditioning response. Pair a stimulus with a reward often enough, and the dopamine response becomes habitual; skip the reward and dopamine neurons shut down. The same thing may happen on the negative side. Dopamine might meditate the learning of "conditioned taste aversions," like Elise's hedonic dysfunction. If punishment follows a stimulus, the animal—including a human one like Elise—learns to avoid that stimulus.

The work of some Italian scientists supports this idea. They trained lab rats to hate saccharine by giving the animals lithium within minutes after the animals drank saccharine-water. (Lithium, I learned, is not high on a rat's list of happy pills, but normal rats love saccharine when they can get it.) The animals learned quickly to avoid saccharine water. But when the scientists first gave rodents a chemical that blocked dopamine's action in certain parts of the brain, the researchers' efforts to train the animals failed. The rats lapped up the sweet water no matter how noxious the lithium.[2]

I wonder how Elise would have fared if she'd had some dopamine blockers in her bag on that fateful trip to Tenerife. Maybe we could be enjoying Malibu together today.

up an example that helps me understand. "Suppose," he says, "you like chocolate. If I give you a lot of chocolate bars and then ask you whether you want another, even though chocolate bars might be your favorite kind of food, at that particular moment you may not want another. You would still like the chocolate, that wouldn't change that much, but you wouldn't want it at that time." Thus, Aldridge concludes, "The liking of rewards and wanting of rewards are two separate psychological components."

In the brain, *wanting*, which he calls incentive motivation, is closely associated with the neurotransmitter dopamine. Block dopamine's action in the brain and animals don't work for a reward anymore. They lose interest in everything, even eating. *Liking* is more closely associated with the brain's natural opiates, the endorphins. Liking sets the pleasure circuits atingle through endorphin action. Some of Berridge's experiments have shown that to be true. "If you inject an opiate-like drug into the brain, you can actually enhance the liking of things," Aldridge says.

In the laboratory, the trick is to isolate single variables so that brain signals for liking, wanting, learning, and sensory processing can be differentiated, one from the other. Aldridge describes a new series of experiments that do just that. His team trained the animals to associate sounds with tastes. One sound told the rats that they were going to get sugar, which they like. A different sound signaled that a strong salt solution was on the way. Just as before, the researchers collected and analyzed recordings from individual neurons in the VP. As expected, the liking neurons fired rapidly in anticipation of sugar. They fired little when the bad taste of salt was expected. "So we were tracking not only the quality of the taste itself, but also the anticipated taste," Aldridge says. Then, as before, the rats were put into a salt-depleted state. In that state, the firings of neurons in their VPs were measured when they heard the sounds alone, without the tastes. Remarkably, the reward neurons now fired rapidly when salt was on the way. That's what Aldridge calls "rock solid data." It shows that the neurons are responding not to taste, but to wanting. The animals don't suddenly like a strong salt solution, but they want the salt.

ELISE'S MALIBU AVERSION

I ask Aldridge about Elise's Malibu aversion, speculating that the experience might have deactivated a hedonic hotspot in her ventral pallidum. It's possible, but Aldridge cautions against jumping to conclusions. "The most one could say based on our findings is that the ventral pallidum might be one place in the brain to look to understand the dysfunction problem," he says. Brain functions are distributed; no single structure—no single neuron—does just one thing. Circuits of neurons in the brain aren't dedicated to a single function. No neurons process pleasure alone. Neurons firing in series aren't fixed like a highway, and they aren't like electrical circuits either. "Information is all over the brain," Aldridge explains, but in experiments like his, it is possible to separate one function from another and investigate the chemical basis of each.

Elise's Malibu malady is, in his view, a completely new question that would take more and different experiments to answer. "We haven't done a test to know what happens when an animal has become sated to a good thing. Maybe cell firing in the ventral pallidum would change; maybe it would stay the same. I don't think the ventral pallidum is the only brain region involved. Learning is part of what controls our desires for food and is likely contributing."

He furrows his brow. "It would take a book to explain how this works in people, and to some degree it would be a work of fiction."

I recall my paperback novel on the beach. Perhaps I'll sip a Malibu after dinner. Elise is traveling on business. She won't be dining with me.

Chapter 13

Cooking Up Some Brain Chemistry

How to cook the perfect steak: If you are grilling outdoors, get the fire as hot as you can. Put the grill down low, close to that hot fire. If you are cooking indoors, put a drip-rack under the broiler and get the broiler unit hot. When you are sure that the grill or broiler rack is very, very hot, plop on the steak and back away. The meat should pop, crackle, spit, and sizzle. If it doesn't, you blew it; your grill or broiler wasn't hot enough.

After about a minute of exposure to intense heat, the cooking side of the steak should have some brown crusts on it. If it doesn't, wait until it does; then turn the steak over and wait for brown crusts to form on the second side. If you like a rare steak, your entrée is now done. If you prefer medium rare or medium (but please, no more than that!), raise the grill surface or lower the broiler pan and turn down the heat. You can slow cook your steak to the desired level of internal pinkness after it is hot-seared. But if you haven't hot-seared it, you haven't grilled a steak; you've steamed a pot roast.

If you are nodding your head in agreement, thinking *Sure, sure, searing forms a layer that traps the juices*, you've got the right idea, but for the wrong reason. The world's leading molecular gastronomist, Hervé This (pronounced tees), has collected all the science that's worth knowing about cooking, and he's come to the conclusion that hot searing does no such thing. For one thing, he says in *Kitchen Mysteries*, all that sizzling and popping are the sounds of liquid—the juices—escaping from the meat and vaporizing into the air. When you plop the grilled steak onto the plate, you see more juices come out, so that brown layer isn't much of a juice trap. In fact, hot-searing contracts the connective tissue that

Do We Eat What's Good for Us?

Your craving for a candy bar probably has more to do with an annoying boss than a nagging neurotransmitter. Still, there's a reason for that sweet tooth that humans share with most other primates. Sweet foods are high-calorie foods, good sources of energy, and important to the survival of wild animals scratching for sustenance in scrub forests and savannas. Natural selection through our evolutionary history favored individuals who tasted the sweet, loved it, and went climbing for more. (In the days of donut shops and ice cream stands on every corner, that survival advantage has dropped from the branches to bite us.)

Today the more salient question is whether our sense of taste prompts us to avoid certain foods that might be unpleasant, harmful, or downright deadly. Here, the answer is clearer, and our perception of bitter and sour tastes offers a real benefit: dangerous alkaloids, strong acids, astringent tannins, terpenes, and saponins are toxic, and their foul tastes keep them out of our mouths and stomachs. Not all toxins taste bad, though, so we rely on learning to teach us what foods to avoid. If a food makes you sick, you probably won't eat it again.

Unfortunately, our protective instincts have their downside. Many healthy vegetables contain bitter compounds. Some of us are less sensitive to them, so we're inclined to relish our broccoli and Brussels sprouts. But for the bitter-taste-sensitives among us, the story is different. Those people miss their recommended "five a day" because they find the bitter taste of many vegetables repugnant.

So significant is the human aversion to bitter (and to a lesser extent, sour) tastes that the food industry has gone all-out to debitter the food supply. Black coffee, green tea, beer, and red wine notwithstanding, you won't find many food products these days that have escaped debittering. Most legumes, fruits, and grains have been bred and cultivated in recent decades to be less bitter, astringent, or sour than were their wild-type progenitors. Orange juice is debittered by the chemical breakdown of a compound called naringin. Most apple juice is treated to remove compounds called polyphenols, which taste bitter or astringent.[1]

When they can, food scientists remove bitter compounds from foods, but that's not always possible. Other approaches to debittering include trapping bitter compounds inside a physical barrier (capsules, coats, emulsions, suspensions) and adding salt and sweeteners to mask bitter flavors. (That's an ancient strategy. Spices and aromatic herbs have been prized for centuries because of their bitter-masking properties, and hot peppers in some form have their devotees worldwide.) Additives that are more modern include molecules that latch onto the offending compounds, tying them up so they become inactive or blocking them from docking with bitter receptors in taste cells.

surrounds the muscle fibers in the beef so much that it probably increases the expulsion of the juices, This contends.[2]

So why is hot-searing the perfect way to cook steak? Because the intense heat on the meat's surface forms a complex cocktail of perhaps six hundred chemical compounds that, together, taste oh-so-good. Although those yummy molecules are sometimes attributed to the Maillard reaction (the same reaction that caramelizes sugar, named after the Frenchman Louis-Camille Maillard), the chemical change-process in steak is actually different. Beefsteak contains hardly any carbohydrate, so there's no sugar to be caramelized. Steak is mostly protein, and scientists say that the searing breaks down muscle proteins—myosin, actin, and myoglobin. The breakdown products combine and recombine in numerous ways, producing such gourmet delights as bis-2-methyl-3-furyl-disulfide, which has a deliciously strong, meaty smell.[3] Now if that's not enough to set your brain's flavor centers a-zinging, I can't imagine what will!

BRAIN A-BUSTLING

The minute that juicy, perfectly cooked steak hits the tongue, taste impulses start traveling to the brain. Where do they go once there? Researchers have in the last decade or so started to use positron emission tomography (PET) and functional magnetic resonance imaging (fMRI) to capture on film those parts of the brain that increase their activity (i.e., receive greater blood flow) in response to taste stimuli. In humans, impulses from the taste buds enter the brainstem and synapse in the region of medulla called the nucleus of the solitary tract. The medulla controls such basic taste-related functions as gagging, swallowing, and vomiting, as well as salivation and the gastrointestinal reflexes that promote digestion. From there, taste signals run a fast track to the brain's central dispatch station, the thalamus, and from there directly to the brain's "conscious of taste" centers.

Place your little finger at a forty-five-degree angle from the outer corner of your eye and you'll get an idea about where those centers are. Japanese scientists demonstrated the precise location when they measured blood flow changes in the brain as volunteers tasted different tastes. The tastes activated the brain's primary taste cortex, which lies in the frontal operculum and adjoining insula. A secondary taste area lies in the caudolateral orbitofrontal cortex, several millimeters in front of the primary taste region. The fMRI scans showed that different tastes are processed in the different part of these regions.[4] Some neurons respond only to sweet taste, others only to umami, and so on.

Oxford psychologist Edmund Rolls has studied the neurons in the orbitofrontal cortex taste areas, where the sensation of flavor comes together or converges. Rolls says that the texture of food can moderate the activity of single neurons in this region. Some of them respond to a food's viscosity, while others respond to the size of the particles in a food or the food's temperature. Rolls studied 112 single neurons in the orbitofrontal cortex that respond to the taste, smell, or sight of food. He found that many are specialists, activated by only one type of sensory input: taste 34 percent, olfactory 13 percent, and visual 21 percent. But other neurons do double duty for two senses; they respond to both taste and sight (13 percent), both taste and smell (13 percent), or both smell and sight (5 percent). But even some of those double-duty neurons have a specialist bent. For example, some that respond best to sweet tastes also respond more strongly to the sight of sweet fruit juice than to the sight of a salt solution.[5]

TASTE, BRAIN, AND BODY

Taste does more than make foods pleasurable. It triggers a series of hormonal changes that affect the body in all its tissues and organs. A sweet taste in the mouth, for example, prompts the pancreas to release two hormones: glucagon, which causes the liver to release glucose; and insulin, which stimulates cells all over the body to take in glucose and burn it as fuel. In a sense, the body has a built-in anticipation response. The sweet taste causes it to "gear up" for the carbohydrate food that is coming.

Taste uses a similar mechanism to prepare the body for protein foods. The trigger for this cascade of changes is the umami taste, which is predominant in the flavoring agent monosodium glutamate (MSG). French researchers didn't find much effect when they fed lab rats water solutions of MSG compared to plain water, but when they added an MSG solution to the animal's ordinary chow, things started to happen. The rats' metabolic rate rose, and the increase lasted a full thirty minutes. The animals' respiratory quotient, a measure of protein utilization, also rose. The umami taste, the researchers concluded, triggered an "anticipatory reflex." The taste initiated changes in metabolism and body chemistry that readied the body to handle the protein foods that were soon to follow—although, in this case, the rat chow was primarily carbohydrate.[6]

The brain and nervous system bring about this response. Japanese researcher Akira Niijima fed solutions of MSG to lab rats. The MSG set branches of the vagus nerve in the stomach, pancreas, and liver humming with impulses. Ordinary salt water had no such effect. Niijima concluded that MSG stimulates taste

I Taste Your Pain!

The experimental designs that scientists come up with are amazing! Here's what some Dutch researchers did. They put some brave volunteers in an fMRI scanner and recorded activity in the brain's IFO (anterior insula and frontal operculum). The investigators already knew that the area steps up its activity when a human being tastes or smells a disgusting substance or views a disgusted facial expression. So they showed their subjects pictures of disgusted, pleased, and neutral facial expressions while the fMRI mapped how IFOs responded. The researchers also gave their subjects a written test of empathy. They studied their results to see how well the test predicted an individual's IFO activity while viewing the facial expressions.

They found that empathy scores predicted IFO activation for both pleasant and disgusted facial expressions. The scientists concluded that the IFO "contributes to empathy by mapping the bodily feelings of others onto the internal bodily states of the observer."[7] In other words, if I share your joy or feel your pain, I do so, in part, with the same region of my brain that despises okra.

receptors in the tongue, which in turn activate the vagus nerve. The vagus nerve then triggers the reflex response that readies the metabolism for incoming proteins.[8]

So the reason that perfectly cooked steak tastes so perfect is the series of neurological and biochemical reactions that begins with the docking of glutamate molecules onto umami receptors in taste cells. The receptor, by the way, is most likely a molecule its discoverers named taste-mGluR4. Its relative, plain-old mGluR4, was first found in the brain, where it influences the production and release of the neurotransmitter glutamate.[9] Yes, glutamate is both a taste agent and a neurotransmitter, which is why the smell and taste of a perfectly grilled steak go . . . well . . . straight to your head.

But maybe all this hoopla about grilling steaks is turning your stomach. *I don't like steak*, you may be thinking, and it's true that some people don't. In fact, you may not prefer protein foods in general. The reason may lie in your perception of the umami taste. British scientists studied the MSG taste thresholds of volunteers. (The taste threshold is the minimum concentration of a substance that a person can perceive.) The researchers wanted to find out if the varying thresholds of individuals predicted the perceived pleasantness of the umami taste or a person's preference for dietary protein over carbohydrate or fat.

As expected, while tasting clear soups containing varying amounts of MSG, some people could taste minuscule amounts, while others couldn't detect MSG at all. The study participants rated how much they liked the soups and how often they thought they might eat them. They also reported their opinions about more than twenty common high-protein, high-carbohydrate, and high-fat food items. On average, the taste threshold failed to predict how well the subjects liked the soups. However, those who could detect small amounts of MSG thought the soup was "meaty." Low thresholds for MSG were also related to "liking" and "preference" scores for protein foods. People who could detect tiny amounts of MSG liked protein foods better than those who couldn't.[10]

Chapter 14

How to Knit a Baby Taste Bud

When I was pregnant many years ago, it was before the era of ultrasound, so I didn't know if the baby was going to be a boy or a girl and I didn't know if the baby was going to be complete, healthy, and whole. My baby girl turned out to be all those things and more. Her development inside me proceeded textbook fashion, far as I could tell, producing all the parts, functions, and senses expected of human newborns.

I wanted to dress my baby in garments I had knit myself, but being science-minded, I didn't spend my entire pregnancy knitting. I read everything about prenatal development I could find during the 284 days I waited for my baby's cry to keep me awake nights. Those pictures in the embryology books fascinated me. What a marvel! From a single cell, baby blood and bone arose. A tiny heart started beating while the embryo looked like nothing more than a tadpole—if I could have looked inside me to see it. Tiny kidneys pinched themselves off, grew efficient little filters, and started making urine. A tiny liver turned red-brown and spongy, getting ready to break down the toxins that would assail her small body once the climate-controlled safety of the womb was abandoned. Inside my uterus, fetal muscles bulged and stretched, fetal intestines contracted and relaxed, fetal nails grew long and sharp. A baby body was knitting itself while my fingers fashioned buntings and bonnets.

My baby knitted together skin for touch, two eyes for vision, two ears for hearing, a nose for smelling, and a cute little pink tongue for tasting. More important, she crafted a brain capable of receiving and processing all those inputs. From the moment of her first breath, she detected and comprehended at some basic level the wavelengths of light and sound; the pressure, cold, heat, and pain of touch; the airborne molecules of odorants; and the chemical stimulants of the five tastes—sweet, sour, bitter, salty, and umami.

Growing a baby is easy. The female body knows how to do it, no tech support required. But for mothers, babies, and the scientists who study them both, the automatic process holds endless fascination, and researchers are only beginning to understand how sensory structures such as taste buds form and how their workings get stitched into the fabric of the brain.

CRAFTING THE SENSE OF TASTE

If I were to write a page of instructions for knitting a baby's sense of taste, I'd have to start with plans for forming the subunits. Needed: one set of taste buds containing all the right cells and receptors; one brain with a region specialized for handling inputs from taste buds; one set of neurons to carry messages from the taste buds to the brain; another set of neurons in the brain to send the appropriate taste signals to other processing centers, such as the emotions where feelings about tastes reside and the cognitive centers where babies make decisions about what they like and don't like, what they'll eat and won't eat.

The diagram for stitching the components of a working sense of taste—or any other sense—begins with cellular differentiation. To stretch the knitting analogy, you need all the right kinds of yarns and stitches in the right places, doing what they should. *Differentiation* in an embryo is the series of chemical and physical reactions that turns cells that are all alike into cells that are different—having different structures and functions and capable of organizing themselves into the body's varying tissues and organs—much as a knitting pattern emerges from single interlocking stitches.

The zygote that became my daughter started dividing soon after sperm met egg. One cell divided into two, two into four, four into eight, and so on, until the embryo formed a hollow ball of cells, all of them more or less the same. That's what she was about the time my home pregnancy test turned blue; but before I could start knitting booties, she had changed.

Somewhere around three weeks after fertilization, the hollow ball of cells, the blastula, altered its shape and structure. During this process, which is called gastrulation, the cells streamed in, over, and around one another, and the hollow ball folded in on itself, creating three cell layers: the endoderm that would go on to become the lungs and digestive organs; the mesoderm, which would form muscles and bones; and the ectoderm, which was fated to develop into skin, brain, and nervous system.

About four weeks after fertilization, I had finished the first pair of booties, and a part of the ectoderm had thickened to form a cell layer called the neural plate. The neural groove ran down the middle of the plate, dividing it into right and left halves. The front part of the neural plate would develop into the brain where sensory processing occurs. The back part would become the spinal cord that carries nerve impulses from the body to the brain.

In the fifth and sixth weeks of pregnancy, I bought yarn for knitting baby blankets and a sweater. That's when the two sides of the neural plate were fusing, forming the neural tube. Three bulges that arose toward the front of the tube developed into the main brain regions: the forebrain, midbrain, and hindbrain. Cell division in the brain was rapid at that time. While I was knitting one stitch every two seconds, neurons were forming at the rate of 250,000 per minute. Each neuron was capable of making as many as ten thousand connections to other neurons in the brain.[1]

As time went on, both my baby and I had a lot more knitting to do. Mine was easy to see and comprehend, but hers was infinitely more subtle and intricate. She needed to fashion all her sensory organs, her spinal cord, and a brain to go with them. Even more important, she had to find a way to stitch the disparate parts together seamlessly to fit and function as a single, integrated sense. Here, I am thinking specifically about taste, so she needed a tongue, several thousand taste buds, sensory neurons, cranial and spinal nerves fibers, and a brain that could let her distinguish five different tastes (and, in conjunction with smell and touch, thousands of different flavors) from no more information that a few molecules docking on some protein receptors in her mouth.

The books I was reading said her taste apparatus developed on a timeline something like this:[2]

- When I was about six weeks pregnant, the first taste papillae started to form toward the back and middle of what would become her tongue.
- Nerve fibers from her developing spinal cord started approaching the papillae when I was about seven weeks pregnant.
- In the eighth week, the fibers began to form synapses with elongated, poorly differentiated taste cells on the tongue's surface.
- By the tenth week, shallow grooves formed where taste buds would eventually develop.
- By the twelfth week, her taste cells were better differentiated, but they still had a way to go, and the number of synapses was nearing its maximum.
- After the fourteenth week, her taste buds were functional. She could probably taste, although what amniotic fluid tastes like we may never know.

NEEDLING THE TASTE BUDS

All this raises some obvious questions. How do the nerves "know" where to go as they grow? How do taste buds and the sensory neurons that carry taste message to the brain "hook up," so that the correct taste nerves get the correct taste impulses and take them to the correct taste regions in the brain?

Sarah Millar, a dermatologist and cell biologist at the University of Pennsylvania School of Medicine, has part of the answer. She and her colleagues have studied the tongue's fungiform taste papillae, the little mushroom-shaped bumps that lie on the tongue's surface and contain a large number of taste buds. Humans have about two hundred of those papillae. (Altogether, several different types of papillae contain a total of about ten thousand taste buds. Each bud contains fifty to one hundred and fifty taste-receptive cells.) Millar has also studied a group of proteins called Wnt. Wnt proteins occur in a lot of places, but wherever they turn up, they seem to do more or less the same thing. They trigger a series of chemical changes that cause cells and tissues to coalesce and shape themselves into organs. Different types of Wnt proteins trigger the development of many ectodermal structures, including hair follicles, mammary glands, teeth, stomach linings—even feathers in birds!

To find out if Wnt plays a role in coaxing baby taste buds to form, Millar zeroed in on one particular Wnt protein, Wnt-β-catenin (β is the Greek letter beta). She genetically engineered some mice to have either a mutant form of Wnt-β-catenin or none at all. She grew their tongue cells in culture dishes. In the mice with the mutant form, taste buds and fungiform papillae grew too large and too numerous. In the mice that lacked Wnt-β-catenin, the taste buds and papillae failed to develop at all.

Millar figured out that Wnt-β-catenin acts as a signaling system. Its production is "switched on" both in taste bud cells and at the places where fungiform papillae eventually form. The series of chemical changes it initiates cause taste buds to develop. Wnt also attracts growing nerve fibers to developing taste cells like ants to a picnic. When the Wnt pathway is working, neurons growing out from the spinal cord find the taste buds with no problem, and the fetal sense of taste is born.[3]

Or almost. As we have to expect, the answer to one question merely raises another question. We have to ask what keeps Wnt from promoting unlimited numbers of taste papillae and unlimited numbers of taste neurons, to the point where the baby ends up being all tongue and tasting-brain and not much more. Wnt has to be controlled in some way. So, something must put the brakes on.

That something—or at least one such something—is another protein called (you gotta love these names) Sonic hedgehog (Shh). A team of collaborators from Mount Sinai Medical Center in New York and the University of Michigan reported, in 2007, still more experiments with cultured tongue cells. They found both Wnt-β-catenin and Shh in developing papillae. Blocking Shh signaling let more Wnt form, and more papillae developed as a result. Adding Shh to the cultures inhibited Wnt-β-catenin signaling and stopped papillae formation.[4]

Nothing is ever simple, so the story goes on, with other regulator proteins getting into the act, among them still more fun names such as Sox 2, Bmp, Noggin, Fgf, and EGF. If the names make you giggle, here's one more: The Wnt-β-

Taste Diminishes with Age

The only thing that we seem to gain with age is weight. Everything else diminishes, and taste cells are no exception. Although taste cells are constantly dying and being renewed (the average life span of a taste cell is about ten days), you'll have fewer at forty than you had at fourteen, which may explain why the older we get, the more we shake on the salt and crushed red peppers.

Older mouths are dryer mouths, too, and less saliva means food molecules don't wash onto the taste buds as well as they once did. Aging often means more trips to the pharmacist than young people have to make, for which those in midlife and beyond are rewarded with medication-induced dry mouth. Dry mouth can also result from smoking, dehydration, radiation treatments for cancer, and too little zinc, copper, or nickel in the diet. And don't forget Sjögren's (pronounced SHOW-grins) syndrome. It's an autoimmune disorder in which the white blood cells, which are supposed to destroy invading microbes, get confused and attack the moisture-producing glands. It is one of the most common autoimmune disorders, affecting some 4 million Americans.

Changes in the mouth may not be the only reason taste declines with age. Changes in the brain may contribute, too. A compound called dynorphin is known to play a part in taste perception. Researchers at the Veterans Affairs Medical Center in Minneapolis studied the genes that direct dynorphin manufacture in the brains of young and old rats. They found less and less activity in several brain regions the older the rats became. The declines occurred in the arcuate nucleus and amygdala, two brain regions strongly associated with feeding behavior.[5] This change may explain, in part, why appetite decreases in the elderly.

catenin pathway is activated when certain genes bind to receptors called . . . are you ready for this? . . . Frizzled.

KNITTING THE TASTE-BRAIN

The taste cells inside taste buds inside papillae are only one part of the story. That baby girl of mine could taste nothing until her infant brain received a traveling nerve impulse. Those neurons that grew toward the taste buds merged into two paired cranial nerves. One is the glossopharyngeal nerve (cranial nerve IX). (*Glosso* means tongue and *pharyn* means throat.) It carries impulses from the back third of the tongue. The other is the chorda tympani branch of the facial nerve (cranial nerve VII). It carries impulses from the front two thirds of the tongue.

Those nerves carry taste impulses into a part of the brainstem called the nucleus of the solitary tract. The solitary tract is the first relay station, but the dispatching there is not simple. Besides acting as the gateway to other brain regions where taste information is received, the solitary tract also connects to several brainstem sites responsible for taste-initiated reflex functions, such as spitting out rotten fruit or rancid fats. Those multiple connections through the solitary tract are especially important for another reason: scientists are still arguing about whether taste delivers signals from the receptors more or less intact and direct to the relevant brain regions or whether signals go through a lot of processing steps before they reach their destination.

Either way, taste messages also travel to the thalamus, which acts as a switching station for sending taste information to various parts of the cortex. Taste impulses also travel to the hypothalamus and amygdala, parts of the limbic system where emotions arise, and to the frontal lobes where decisions such as "Hmmm! I love breast milk" are made in newborn brains. So knitting the baby taste bud is only one small stitch for woman- and baby-kind. It's knitting the baby brain that matters.

Chapter 15

Expecting What You Taste and Tasting What You Expect

I once prepared and served a meal that was nothing but Jell-O. That's not quite true. There were other things mixed in with the Jell-O, but every dish in the meal had Jell-O at its heart.

The occasion was festive, long promised, and much anticipated. It came about because an out-of-town friend gave me a cookbook of Jell-O recipes as a joke. The joke was more on him than me, because he is a self-confessed Jell-O freak. He loves Jell-O. Being from Minnesota (no insult intended to the good people of Minnesota), my friend dislikes spicy foods, anything green, anything fresh, and most things that have any texture, color, or flavor. He describes himself as "a Velveeta cheese and tater tots kinda guy." He admits that he'd be more than happy to eat Jell-O thrice daily. Thus the cookbook and thus my promise to him: When you visit next year, I'll make you a meal that's entirely Jell-o.

And make that meal I did. We sat down to a Jell-O feast. Our asparagus soufflé salad was a bit of a cheat—it was made with unflavored gelatin—but the rest of the menu was the real deal. I served avocado-horseradish aspic (in raspberry Jell-O), tuna salad (in lime Jell-O), meatloaf (no-bake, made with lemon Jell-O), and banana bread (made with apricot Jell-O). For dessert, we ate strawberry pie, featuring my friend's favorite color and flavor of Jell-O. We had coffee-flavored Jell-O and cream with our brandy to end with a touch of class.

The reviews were mixed. The guest of honor was in Jell-O heaven, though he passed on the aspic. (Too spicy!) Personally, I more or less liked it all, but my favorite was the meatloaf. Everybody else ate the requisite amounts of each dish and pronounced the meal successful, although perhaps not memorable. Jell-O,

Tasting What You Know and Knowing What You Taste

Nothing's wrong with your sense of taste, right? Think again. Researchers say we are probably pretty good at recognizing tastes when things are working right, but when we lose taste perceptions, we're clueless. Here's how they know: A research team at the University of Pennsylvania School of Medicine asked nearly five hundred volunteers (mean age fifty-four) to respond "easily," "somewhat," or "not at all" to such statements as, "I can detect salt in chips, pretzels, or salted nuts," "I can detect sourness in vinegar, pickles, or lemon," "I can detect sweetness in soda, cookies, or ice cream," and "I can detect bitterness in coffee, beer, or tonic water." Then the researchers gave the subjects various foods to taste. Those subjects who tasted normally knew it, but those with taste dysfunctions weren't aware of their deficit. Ruling out smell as a variable using other tests, the study showed that ageusic subjects think they can taste what they don't. Sex and age make a difference. Older adults get it wrong more often than younger ones; men generally do worse than women.[1]

many declared, just wasn't their favorite dish, although the Jell-O was hardly discernible in anything but the tuna salad.

I have to wonder why they aren't still raving all over town about that culinary extravaganza. I think I have an answer: It was just the *idea* of all that Jell-O! Taste is not so much what we eat as what we *expect* to eat. Taste is only partly palate. It's more about the brain, and the brain gets what it thinks it will.

ANTICIPATION

Anticipation is a powerful engine. It drives the placebo effect (see Chapter 4), and it has more to do with taste than even a cordon bleu chef might predict. Certain regions of the brain become active when a pleasant sensory input is expected. That activation then moderates the activity of sensory processing areas —so much so that what we taste is more about what we expect than about what we actually get.

It's not merely a matter of visual impact, either—not simply color, anyway. Although we expect yellow pudding to be lemon and purple juice to be grape, color alone doesn't set up much in the way of anticipation. A number of researchers have tested color and flavor associations, often using preparations of the same sweetness but of different colors. Results are usually unremarkable, except that red is often perceived as the sweetest and blue as the least sweet.[2] Liquids are rated sweeter than solids, and colored solutions are considered sweeter than colorless ones.[3] People typically associate yellow and orange with citrus flavors; they associate red, yellow, and orange with fruity flavors. Generally brown, gray, black, and blue don't fare well, no matter what the taste descriptor.[4] The influence of odor on taste is probably more important than that of color. Participants in one study rated a strawberry-scented red solution as sweeter than an unscented one of the same hue, although there was no actual difference in the sugar content.[5]

Color may not influence taste as much as it does the broader, overall perception of whether a food is appetizing or even marginally palatable. I remember in college how everyone complained (and worried) about the turquoise-blue tinge detectable on the margins of the meat served in the dormitory cafeteria. I was studying science, so I knew that the color was the result of preservatives that wouldn't hurt us unless we ate them by the truckload. But my roommates only eyed me with distrust when I offered rational reassurances. There's something wrong with blue meat! Any fool can see that! The conclusion was not that the meat tasted bad, but simply that it should not be consumed at all.

Experience plays a role, too. Whether you taste something now depends on whether you have tasted it before—and whether you continue to taste it. Many

We Know What We Know

When it comes to taste, we know what we know and we aren't about to be persuaded otherwise. A team of researchers from the Universities of Washington and Florida gave test subjects a mild and creamy cheese with an orange rind and a stronger tasting, dry cheese with a purple rind. They then had the volunteers rate the cheeses on a scale from mild to strong flavor. The participants rated the orange-rind cheese as quite mild.

Later, the subjects rated other cheeses for creaminess, not flavor. Their past experience might have taught them to associate the orange rind with creaminess, but it had not. Their "training" had taught them that orange rind meant mild flavor. Creaminess simply wasn't in the equation. "Once people learned that a cue predicted an outcome, they became less likely to learn about this very same cue with respect to a different outcome," the investigators wrote.[6]

This finding is more than a curiosity. It suggests that once we've learned one taste association, we are disinclined to learn another. If, for example, you learned as a child that green vegetables taste bitter, you may have trouble accepting them into your adult diet despite their numerous health benefits.

Can You Imagine a Flavor?

If we taste what we expect to, can we imagine a taste? Maybe. Researchers at Emory University in Atlanta showed people pictures of foods and pictures of scenery while fMRI scans were taken of the volunteers' brains. The scenery pictures didn't activate their primary brain taste centers, but the food pictures did.[7]

The mere thought of food may be enough to get the mental juices flowing. Scientists in Japan used fMRI to scan the brain activity patterns of subjects who did nothing more than think about pickled plums (*umeboshi*), a traditional Japanese food with a strong sour flavor. The scans showed a pattern very similar to that seen when the pickle was actually tasted.[8]

people can't taste MSG the first time they try it, but after some training, they can.[9] The same thing happens with sugar, although you won't find many people unfamiliar with its taste. Nonetheless, when Massachusetts researchers "trained" their volunteers with the sugar fructose, sensitivity to the taste of another sugar, glucose, rose. The effect lasted for about eleven days, but it was a "use it or lose it" thing. The effect disappeared after a month.[10]

The training effect shows up in more than taste tests. French scientists captured fMRI images from volunteers who tasted harmless chemicals they had never tasted before. Over several weeks, as the subjects' ability to detect the new flavors increased, so did their brain activity in certain regions.[11]

TASTE AND THE BRAIN

At least one neurotransmitter, serotonin, is known to carry messages between neurons in two sites: in the taste buds of the mouth and in the brain. In the taste buds, impulse-generating cells release serotonin after tastants bind to nearby taste-receptor cells; the release starts the "taste message" traveling toward the brain. In the brain, serotonin sends all kinds of signals from one brain cell to another.

Thus, it's not hard to see a link between the sense of taste and mental health, and researchers indeed find such links. Deficiencies in serotonin and other neurotransmitters bring on mood disorders, which often carry with them a blunting of the sense of taste. Depression, for example, decreases sensitivity to all tastes, especially sweet. Panic disorders reduce sensitivity to bitter, and stress induced in normal individuals increases sensitivity to the bitter taste of saccharin.[12] This association suggests that reduced levels of particular neurotransmitters may account for both certain mental complaints and the dysgeusia, or impairment of the sense of taste, that goes along with them.

To see whether there is anything to that idea, researchers at the University of Bristol in the United Kingdom tested twenty healthy volunteers for their anxiety levels. The most anxious participants were less sensitive to bitter and salty tastes. Then the scientists gave all the subjects two antidepressant drugs. One raised the level of the neurotransmitter serotonin the brain; it also increased people's sensitivity to sweet and bitter tastes. The other drug increased the amount of another neurotransmitter, noradrenalin; the result was greater recognition of bitter and sour tastes.[13]

It doesn't take a drug or chemical to change expectations and the tastes that result from them. A team of University of Wisconsin-Madison scientists captured images of the brain fooling itself. Psychiatrist Jack Nitschke recruited volunteers to have MRI pictures of their brains taken while they tasted solutions of bitter quinine, sweet sugar water, or tasteless distilled water. The researchers and the subjects agreed ahead of time on a set of clues to precede the solutions. A minus

Craving Chocolate

If you want to understand why you crave (or don't crave) chocolate, put your brain inside an fMRI scanner and take a peek at what's going on. That's what Oxford researchers did with sixteen women. Eight of them craved chocolate; eight of them didn't. All were of normal weight, although the cravers consumed, on average, a whopping 370 grams (thirteen ounces, or nearly nine bars)) of chocolate weekly compared to 22 grams (less than one ounce) a week for the noncravers. The researchers studied the women's brains while they responded to the sight of chocolate, the flavor of chocolate, or both together.

The cravers and noncravers showed no differences in how their brains responded in the primary taste- and visual-processing regions of the brain, but both the sight of chocolate and the flavor of chocolate produced greater activation in the orbitofrontal cortex and ventral striatum of cravers compared to noncravers. Sight combined with flavor produced a response that was "greater than the sum of its parts" in the orbitofrontal cortex and cingulate cortex of cravers. The greater the activity in those regions, the more pleasant the women rated the chocolate.[14]

Among other things, the orbitofrontal cortex is a major center for goal-directed behavior. It sends us toward the desirable and away from the undesirable. Stimulus/reward associations activate the ventral striatum. The cingulate cortex has many functions, among them the linking of cognition with emotions.

sign flashed through fiber-optic goggles signaled quinine coming next. A zero signaled the tasteless distilled water, while a plus sign said a sweet treat was on the way. During the experiment, however, Nitschke randomly mixed the tastes and the cues, so that the signal did not necessarily precede its assigned taste. The interesting outcome occurred when a plus sign preceded the bitter quinine. The brain responded in a characteristically "bitter taste" pattern, but less strongly than when the expectancy sign was zero or minus. In other words, if the brain expected a sweet treat, the bitter medicine tasted less bitter.[15]

THE POWER OF WORDS

Words can do the trick as well. Researchers at Oxford dripped solutions of MSG into the mouths of volunteers whose brains were being scanned with fMRI. The volunteers also rated the solutions according to intensity and pleasantness. Their primary taste cortices got it right when it came to intensity. The participants could judge accurately which solutions contained more or less MSG, and they found the highest concentrations both intense and unpleasant when tasted in a neutral environment.

The malleable factor was pleasantness. When the words *rich and delicious taste* preceded administration of the solution, people judged the solution as more pleasant than when the words *monosodium glutamate* appeared on a projection screen. Pleasantness was perceived as greater also for the words *rich and delicious flavor* compared to *boiled vegetable water.* The MRI scans showed where those responses originated. There was no difference in activity of the primary taste cortex. The differences lay in the secondary processing regions of the orbitofrontal cortex and cingulate cortex. Those are the areas of the brain that represent the emotional value of sensory rewards. The brain's ventral striatum became active in response to the word labels, too. It is a region of automatic learning, where we learn to associate a stimulus with a reward.[16] What's more, paying attention makes a difference. When people are told to remember and rate the pleasantness of a taste, brain activity in those regions increases.[17]

If that's true, then how much of the flavor of that Jell-O dinner came from the dishes themselves and how much came merely from the word *Jell-O?* How many of the so-so reviews came from the food and how many came from obsessing about *all that Jell-O?* In the interests of science, I'm planning another Jell-O dinner, but this time I'm not telling anyone what they're eating. They'll be raving about that meal for decades.

PART FOUR

VISION

Anyone who has common sense will remember that the bewilderments of the eyes are of two kinds, and arise from two causes, either from coming out of the light or from going into the light, which is true of the mind's eye, quite as much as of the bodily eye.

—Plato
The Republic

Chapter 16

The Big Picture

Vision is the best understood of the senses. We know a lot about how it works, in part because we can see what happens when people lose some component of their visual sense.

CASE STUDIES[1]

Harry W. suffers from akinetopsia, or motion blindness. He sees no continuous motion. He perceives, instead, what he describes as a series of still images with blurs in between. Harry, whose motion blindness has worsened along with his Alzheimer's, likes to watch his cat race across the living room carpet. He sees no motion as the cat bounds and leaps, but he knows the cat is moving because the animal's form appears at different places in the room. It's only when the cat curls up to sleep on a sunny windowsill that Harry sees his pet clearly.

Marianne R. has no trouble with motion vision but she cannot discriminate objects. Her condition is called visual agnosia. It struck Marianne after a massive stroke damaged parts of both her occipital and temporal lobes. When working with her occupational therapist, Marianne can pick up an apple, peel it, and eat it. She can describe its color as red, its shape as spherical, and its texture as waxy, but she cannot say that the apple is an apple. She cannot identify geometric shapes, the letters of the alphabet, or a photograph of a dog as a dog.

Alex S. is only eleven years old, but he has Balint's syndrome, probably as a result of a massive infection he incurred in early childhood. The infection caused bleeding in his brain, and blood pockets had to be drained to relieve the pressure. Now, Alex has all three of the conditions that define Balint's. He can't see or perceive several objects at the same time, a disorder called asimultagnosia. When he looks out his window at the city park near his home, he sees a single tree, but cannot discern the bike path in front of the tree or the cyclist pedaling along the path. At the mall, he bumps into people not because he is careless, but because he doesn't know they are standing in his path. Although his visual acuity tests are normal, he has trouble picking up objects (optical ataxia) and directing his gaze toward a single object (optical apraxia). He does poorly in school. In order to read, he has to move the book from side to side to see the page. He skips letters and words and cannot follow lines of text.

Harry, Marianne, and Alex demonstrate a key feature of visual processing in the brain: individual capacities can be lost, while others are retained. This occurs because parts of the brain's visual-processing machinery can operate independently. They specialize—handling motion, orientation, color, or other specifics. Other regions of the brain, beginning as early as the retina of the eye, "put it all together" to achieve what the higher brain centers perceive as a unified image of the world. The story begins in the eye.

THE RETINA

The retina lies at the back of the eye, but it is actually part of the brain. A mere half-millimeter thick,[2] the retina is made up of three parallel layers of nerve cells, with their tangled axons and dendrites forming alternating rows in between. The layer closest to the front of the eye consists primarily of ganglion cells. The second layer contains horizontal cells, bipolar cells, and amacrine cells. In the deepest layer at the back of the retina lie the light-sensitive neurons—the photoreceptors, cones, and rods—all 120 million of them.[3]

Cones are responsible for color vision in bright light. A single cone is individually fine-tuned to respond only to blue, red, or green light. In the center of the retina lies the macula, the region that facilitates sharp vision in the middle of the visual field. A spot in the macula called the fovea is composed almost entirely of cones; it's where the sharpest images focus in bright light. Rods, which respond to shades of gray, work best in dim light. Rods are plentiful in the peripheral areas of the retina, outside the macula. The fovea contains no rods.

The responses of rods are slow compared to those of the cones, but in the end, the result is the same. When light strikes a rod or a cone, a specific chemical reaction occurs, triggering a nerve impulse, and messages travel among neurons in the eye, although not always in the usual fashion. Most neurons in other parts of the body transmit signals across the synaptic gap (the space between neurons) via the chemical action of a neurotransmitter. That does not always happen in the retina. Some of the signal transmission in the eye is by direct, electrical means. Still, a large number of neurotransmitters operate at various stages of retinal processing, sometimes transmitting messages and sometimes influencing the flow of electrical signals.

One important neurotransmitter at work in the retina is glutamate, but it plays a kind of backwards role. In the dark, photoreceptors constantly emit glutamate. The bipolar and ganglion cells that fire in response to glutamate control the OFF pathway in retinal processing. That pathway handles dark images viewed against a light background. When photoreceptors are exposed to light, an impulse is triggered and glutamate release ceases. Certain bipolar and ganglion cells stop firing when glutamate is absent. Those cells control the ON pathway, which allows detection of light images on a dark background.

Individual rods and cones in the deepest layer of the retina have very narrow receptive fields; that is, each responds only to a single, minute point of light that strikes it directly. The signal a rod or cone generates travels outward (toward the front of the eye) into cells in other layers that have larger fields. In the surface layer of the retina lie the ganglion cells. These cells have wide, circular receptive fields. Ganglion cells are of two main types. Some respond most to light in the center of the circle. Others are activated when light falls on the periphery. Together, these ganglion cells collect and organize impulses from the cell layers below. They are responsible for most of the signals that travel to the brain via more than a million optic nerve fibers.[4]

Pathways from the fovea to the ganglion cells consist of a single cone that sends an impulse to a single bipolar cell and from there to only two ganglion cells: one of the ON type and one of the OFF type. Communicating with both ON and OFF cells means that the message from the fovea is not lost whatever the object-background relationship. The straight-line route means that information from the fovea travels to the brain virtually unchanged. The blue-detecting cones (found outside the fovea) operate a little differently from the red and green; they transmit signals through different types of bipolar and ganglion cells. The ON response is blue, and the OFF response is yellow.

Rod-dominated vision in dim light works through slightly different pathways. The bipolar cells that receive messages from rods do not connect directly with ganglion cells. Instead, amacrine cells in the middle layer of the retina act as go-betweens, ferrying signals along to both ON and OFF ganglion cells, and from there to the brain. It was once thought that only rods sent messages via amacrine cells, but recent research suggests that at least some cones may transmit via amacrine cells, too.[5]

IN THE BRAIN PROPER

Some of the impulses that travel from the ganglion cells of the retina through the optic nerve never reach the brain's visual centers. Some follow neuronal branches to other structures in the brain, including the suprachiasmatic nucleus (SCN), which is the body's primary time-keeping mechanism (see Chapter 30). Some impulses go to parts of the brain that control automatic movements, such as the dilation and constriction of the pupil in responses to changes in light level. Some signals travel to a region called the superior colliculus. It triggers the automatic movements of the eyes and head needed to keep a sharp, central image focused on the fovea. It also selects from the complicated visual environment the specific object worthy or focused visual attention.[6]

But the first stop for 80 percent of the impulses that travel through the optic nerve is the lateral geniculate nucleus (LGN), which lies in the thalamus of the brain. The thalamus acts as a "central dispatcher" for the information that comes in from all the senses except smell, but the thalamus is more than a sorting station. Certain structures in the thalamus initiate and sustain visual attention. Another part of the thalamus adjusts the brain's perception of orientation, so that an image appears stable no matter what the position of the head and body.

The LGN is organized into six cell layers. The layers have distinct functions. In adults, each layer of the LGN receives inputs from one eye only. (This specialization develops through stimulation of both eyes early in infancy; it's the reason why a newborn's eyes should never be covered.) In the LGN, some crossover occurs. Information from the right visual field goes to the left hemisphere of the brain and vice versa. These signals end up on the two sides of the occipital lobe (at the back of the head) in the primary visual cortex. That area is called the striate cortex or V1.

V1 is organized in horizontal layers much the same as the LGN. The layers have their own visual specialties. Some respond only to a certain color or to a movement in one direction. Some detect only pinpoints of light; others, only edges and lines. So precise is their action that one line-detecting neuron "sees" only horizontal lines, while another detects only right angles.

The horizontal layers of the V1 are also organized into vertical columns. Each column contains all the layers but handles a separate segment of the visual field. V1 gets (more or less; see Chapter 18) an exact neural replica of the light patterns that fall on the retina. Subsequently, higher visual-processing centers can draw upon and utilize the visual information coded in the V1.

Nearly thirty different visual-processing regions in the brain have been named and described. The best known are the primary area, V1; the secondary area, V2; and the so-called higher associative areas, V3, V4, the medial temporal (MT) cortex (V5), and others, up to V8. Some of the specialties of these areas are known. Like V1, V2 maintains a complete and accurate map of information coming in from the retina. V3 is finely tuned to changes in orientation. It also seems to "understand" the apparent change in shape that occurs when an object is tilted; it "knows" the object still has the same shape, although the point of view has

Wiring the Brain for Vision

The brain isn't always in charge of the senses. If fact, there's reason to argue the opposite: the senses run the brain. That's certainly true when we consider how the visual system develops in infancy. As with every other sensory system, the control mechanisms are chemical. Scientists at Children's Hospital in Boston have found (in mice) a protein called Otx2. It's made during critical periods when the visual system is developing in the brain, but the brain doesn't make it. The retina does, but only when the eyes are getting light stimulation. Otx2 travels from the eyes to the brain where it triggers the wiring of the brain's visual circuitry.[7] The process has to be timed to perfection. "The eye is telling the brain when to become plastic, rather than the brain developing on its own clock," says researcher Takao Hensch. "If the timing is off, the brain won't set up its circuits properly."[8]

changed. V3 sends impulses to both V4 and MT (V5). V4 neurons are fine-tuned for shape, orientation, and color.

The major motion-detection area is MT (V5). It receives input from V1, V2, V3, and V4. No matter what the object, if it moves, MT neurons fire. Neurons in the MT are organized in columns and are tuned specifically to particular line orientation, much as in the LGN. MT also receives inputs directly from the retina, with no stopovers in V1 through V4. Thus, doctors sometimes encounter unusual cases in which a tumor, stroke, or injury in V1 renders a patient blind to visual images but still capable of detecting motion.[9]

MT isn't the only motion-processing region of the brain. The medial superior temporal (MST) cortex responds to motion in a straight line, but its neurons are also tuned to circular motion (clockwise or counterclockwise) and radial motion (expansion or contraction toward or away from the viewer). They also handle the visual inputs that come from a complex series of movements, such as when crossing a busy street. The superior temporal polysensory (STP) area specializes in biological movement. It lets the human brain recognize another human being from only a few movement clues.

Other associative regions that process images lie outside the main visual areas. The inferotemporal (IT) cortex, for example, receives many impulses from V4. It responds to both colors and shapes and is thought to play a part in recognizing objects and remembering images. One researcher, Nancy Kanwisher at the Massachusetts Institute of Technology, has used functional magnetic resonance imaging (fMRI) over more than a decade to locate and describe a number of other associative areas that she considers "functionally distinct." They include the fusiform face area (FFA), which responds to faces; the parahippocampal place area (PPA), which responds to places; and the extrastriate body area (EBA), which responds selectively to images of human bodies or body parts. "The selective responses of these regions are robust enough that each of them can be found, in the same approximate anatomical location, in virtually every normal subject scanned with fMRI. Thus, the FFA, PPA, and EBA are part of the basic functional architecture of human extrastriate cortex," Kanwisher says.[10]

Other researchers have identified additional areas that perform unique visual functions. One area at the front of the brain just above the eyes specializes in the faces of infants. It responds immediately (100 milliseconds) and instinctively to the faces of the very young, but not to more mature faces. It's an emotional area of the brain, and its selective responses to baby faces may play an important role in bonding adults to the infants in their care.[11]

The newest research also shows that visual processing isn't limited to the neurons of the brain's outer layer, the cortex. In fact, most of the brain's cells are not neurons. They are glial cells. Somewhere between one and five trillion glial cells protect and support the neurons of the two hemispheres.[12] One important type of glial cell is the astrocytes, or astroglia. Scientists once believed that astroglia merely provide nutrients, support, and insulation for neurons, but recent research suggests that astroglia do much more. They "instruct" new, unspecialized brain cells, "teaching" them what kind of mature neuron to become. They also help regulate the transmission of nerve impulses and the formation of connections between brain cells.[13] In the visual system, they respond directly to visual stimuli much as nearby visual neurons do—sometimes even more finely tuned to specific features, such as orientation, than neurons are.[14]

Different pathways and connections between and among cell types and regions take separate tracks in the brain. One major route is thought to head off to the temporal lobes where objects are recognized and identified. A second major pathway leads to the parietal lobe where the size, position, and orientation of objects are processed. This is the kind of information the brain uses not to name objects but to manipulate them. These two separate, but interrelated, tracks have been the subject of a great deal of research. Scientists at Ben-Gurion University in Israel, for example, presented volunteers with optical illusions in which one small bar appeared longer than another, when in fact the opposite was true. Participants also saw bars of equal length and bars in which the longer-looking one was actually longer. In each case, the volunteers visually estimated and reported bar lengths. Then, they either grasped the bars between thumb and forefinger or indicated the lengths of the bars by holding their thumb and forefinger apart at a distance they judged equal to the lengths of the bars. In nearly every case, visual perception was fooled, but the action sense was not. Even when participants thought the shorter bar was longer, their fingers, nonetheless, got the length comparison right.[15]

When Sight Is Gone

If you lose your sight, do your other senses sharpen? That question has been much debated, but some evidence suggests the answer is yes. Furthermore, the adaptation may occur in a matter of days.

Researchers at Beth Israel Deaconess Medical Center in Boston have found that people with normal vision who are blindfolded for five days learn Braille better than people who aren't blindfolded. Brain scans of the blindfolded Braille students reveal that the brain's visual cortex becomes extremely active in response to touch. No such activity is observed in pretest scans, and the increased activity disappears within twenty-four hours after the blindfolds come off.[16]

"The speed and dynamic nature of the changes we observed suggest that rather than establishing new nerve connections—which would take a long time—the visual cortex is unveiling abilities that are normally concealed when sight is intact," says senior author Alvaro Pascual-Leone. Says principal investigator Lotfi Merabet, "In a sense, by masking the eyes, we unmask the brain's compensatory potential."[17]

CROSSTALK, ADAPTATION, AND PLASTICITY

Harry, Marianne, and Alex experience visual difficulties because of damage to a small region of the brain's visual-processing centers, somewhere within the V1–V8 network. Each one of them has a lesion in a different area, so their losses in function are different. The localized damage robs them of a particular ability, while leaving the remainder of their visual capacity unimpaired. For each of them, it's possible that the brain's plasticity will allow a network of neurons nearby to take over and perform the function that was lost.

All three are enrolled in experimental treatment programs designed to help them overcome—or at least lessen—their visual deficits. All have experienced improvement for three fundamental reasons. One is that different regions of the visual system "talk" to one another, and not always in a bottom-up, direct-from-V1 fashion. Considerable feedback travels from the so-called "higher" regions back down to those areas that handle basic processing.

This leads to the second reason: adaptation. Sensory neurons change their responses to match changes in input. As amazing as that sounds, neurons operate differently depending on what the environment is serving up at the moment. Adaptation modifies how information is coded in populations of neurons. Studies in monkeys, for example, have shown that neurons of the V1 quickly find an efficient way to handle input from a fixed visual stimulus. They grow more efficient, more organized, and less variable in their activity.[18]

The third reason is plasticity, the brain's ability to reorganize itself. Neuroscientists once thought that the adult brain was static, with its only change being the inevitable, day-to-day loss of neurons. But research evidence that began accumulating in the 1990s proved that idea wrong. Faced with a lesion, the brain can sometimes reroute sensory input into alternative processing areas. When existing connections among neurons are lost, new ones can be forged. New neurons can even be born, differentiated, and remodeled to take over lost functions.

Those processes are aided by experience. Alex is now performing better in school because he has been doing the same kinds of training exercises that educators recommend for children with dyslexia. Harry will probably never perceive his cat's motion, but he can navigate his home, using landmarks as clues to the direction of his own motion. As for Marianne, she smiled at her therapist last week when she called an apple an apple.

Chapter 17

Color and Memory

Want to remember a picture or scene? See it in living color! So says a team of psychologists working at the Max Planck Institute in Germany. They showed pictures of forests, rocks, and flowers to volunteers who looked at a large number of photographs, half in color and half in black and white. Then the volunteers viewed the same images randomly mixed with many new images. Their task was to identify which pictures they had seen before and which they had not. Memories of the colored pictures were 5 to 10 percent better than those of the black-and-white ones, no matter how long the subjects viewed the photos. Subjects who first saw pictures in color and later saw them in shades of gray (or vice versa) did not remember the pictures as well.

Through several related experiments, the researchers were able to rule out the idea that their volunteers were simply paying more attention to the colored pictures. The investigators also found that artificially colored pictures, as opposed to natural colors, offered no memory advantage.[1] The study concluded that the human brain stores colors as part of the overall memory of scenes and objects. The question is how?

FIRST THINGS FIRST: HOW COLOR VISION WORKS

A physicist will tell you that there is no such thing as color in the real world. Like beauty, color is in the eye—or more accurately, the brain—of the beholder. Light acts like particles, but it also behaves like a wave. The distance between the peaks of the waves can vary. That's wavelength. White light contains many wavelengths. Objects reflect different wavelengths, but wavelengths themselves have no color. Color vision is just the brain's way of telling one wavelength from another. Humans perceive wavelengths between 400 and 700 nanometers (nm) long as colors ranging from violet (the shorter wavelengths) to red (the longer ones). Within that range lie many thousands of color variations that the human eye—and brain—can differentiate.

The color-sensitive cells of the retina, the cones, come in three varieties: the S- or short-wavelength cones that are most sensitive to wavelengths around 430 nm; the M- or medium-wavelength cones that respond best around 530 nm; and the L- or long-wavelength cones that have their optimal response at 560 nm.[2] These cones are often dubbed the blue (S), green (M), and red (L) photoreceptors, although their color responsiveness is not, in truth, that limited. Individual cones respond to light across a wider range of wavelengths; their designation as S, M, or L comes only from the wavelength that elicits the strongest response. The three cone types vary in their placement in the retina. There are no S-cones in the fovea. Outside the fovea lie about twice as many L-cones as M-cones, but their arrangement is not uniform or predictable; cones of a single type can occur in clusters.

Each of the cone types contains a photopigment that changes chemically when struck by light. Photopigments have two main chemical components: a chromophore, which, in humans, is retinal (a form of vitamin A); and opsin, a protein. Although each opsin has a different molecular structure depending on the cone type, the function of S-opsin, M-opsin, and L-opsin is the same in all cones (and the same as that of rhodopsin, the protein in the black-and-white photoreceptors, the rods). When light strikes the photopigment, the opsin and the chromophore split apart. That chemical change triggers a nerve impulse.

The impulse from a cone travels through a bipolar cell in the retina's middle later to a ganglion cell in the outer layer. These "spectrally opponent neurons" are excited by one cone type and inhibited by another. For example, a signal from an L-cone excites a +L/–M cell, but a signal from an M-cone inhibits it. Conversely, +M/–L cells are excited by inputs from M-cones and inhibited by inputs from L-cones. Signals from the S-cones work in opposition to both L- and M-cone signals. For example, +S/–(L+M) cells fire in response to short-wavelength light; medium- and long-wavelength light signals block their firing.

The opponent process theory of color vision suggests that these cells "recode" information from the cones into three processes or channels: one no-color channel (white-black, or the "luminance" channel) and two opposing color channels (red-green and blue-yellow). The new channels arise from differing patterns of signals from the three cone types. The process uses, in theory, simple addition and subtraction:

> In the L + M or luminance channel, the signals from L- and M-cones are added to compute the intensity of the stimulus. In the L - M colour-

> opponent channel, the signals from L- and M-cones are subtracted from each other to compute the red-green component of a stimulus. Finally, in the S – (L + M) channel the sum of the L- and M-cone signals is subtracted from the S-cone signal to compute the blue-yellow variation of a stimulus.[3]

This reorganization of color information into three channels occurs both in the retina and in three layers of the lateral geniculate nucleus (LGN) of the brain. Cells in the LGN specialize in the inputs they receive, keeping the "channels" organized pretty much the same way as the ganglion cells do. The magnocellular layer carries luminance information, and motion, too. The koniocellular layer handles the blue-yellow. The parvocellular layer carries the red-green channel. Damage to the parvocellular layer results in an extreme loss of color vision.

Another way to look at pathways is to examine them from the ON–OFF point of view. ON and OFF describe how luminance affects a neuron's firing, regardless of wavelength. In color vision, ON means a neuron's firing rate speeds up when illumination *increases*. OFF means the firing rate speeds up as illumination *decreases*. The majority of S-cone input is of the ON type. A specialized kind of bipolar cell in the retina's middle layer provides S-ON signals to the ganglion cells and LGN. Three types of ganglion cells in the retina receive S-cone input: one type receives excitatory input, and two types receive inhibitory input. The main conduit for S-ON signals in the LGN is the koniocellular layer. The parvocellular layer in the LGN receives little or no input from S-cones; some cells in the magnocellular pathway may respond to S-cone signals, although their sensitivity to L- and M-cone signals is much greater.[4]

From the LGN, visual information travels to the primary visual cortex, or V1, in the occipital lobe (at the back of the head). Most V1 cells are more "interested" in the orientation of edges and differences in illumination than they are in color, so they don't show the same pattern of color activation that's seen in the LGN. Only 5 to 10 percent of V1 neurons respond strongly to color alone (red-green and blue-yellow), but as many as half respond to some combination of color with luminance.[5] The color-sensitivities of V1 cells differ markedly from those of their counterparts in the LGN, so investigators assume that some recombination of color information is occurring as signals travel to the V1 or in the V1 itself.[6]

Although the process is little understood, the information about color that leaves the V1 and travels to other visual-processing regions of the brain must be separated from other types of information, such as the shape, texture, and form of objects. If it were not, you would be unable to see objects in black and white. Yet somehow, as the experiment on better memory for naturally colored scenes suggests, color information also stays linked to other attributes of an object. Thus, you can perceive an apple in shades of red or shades of gray, but if you see the red one, the red information stays linked to the information that was also available in black and white: the size, form, and surface characteristics of the apple. Neurons capable of achieving this integration are known to lie in V2; the color-sensitivity of those cells depends on the background. About one third of the cells in V2 are narrowly tuned to a small range of wavelengths; those cells could be important to recognizing separate objects in scenes and detecting color in shadowed or dappled environments.

At least two higher color-processing areas (above V1 and V2) have been identified and studied. Researchers can't always agree on the brain's topography or the names given to its regions, so the areas known to be especially color-responsive go by various designations, usually V4 (also V8 or VO-1) and V4-alpha. Damage to V4 sometimes induces a form of colorblindness called cerebral achromatopsia, in which a stroke or trauma victim sees the world in black and white, while the perception of form, depth, and motion remains unimpaired. V4-alpha is active during tasks requiring color ordering, color imagery, knowledge about color, color illusions, and processing of object color.[7]

A number of experiments have demonstrated the color-affinity of V4 and V4-alpha. In one, researchers at Baylor College of Medicine in Houston used fMRI to pinpoint neurons near the V4 that fired in response to the color blue-purple. Subsequent electrical stimulation of those same neurons through tiny electrodes caused a volunteer to "see" blue-purple, although no color was actually shown. "[It's a] blue, purple color, like aluminum foil when it burns," the man reported. In response to stimulation of neurons in the V1, he "saw" only flashes of white light.[8]

Yet, as much as our sense of order might demand that we identify specific brain regions for specific types of processing, nature refuses to cooperate. Evidence is accumulating that color analysis and coding are inextricably linked with the analysis and coding of other visual elements, such as motion and form. For example, difficulty in recognizing shapes often accompanies achromatopsia, suggesting that color perception and shape recognition may be handled in the same brain regions. Experiments show also that V4 qualifies as the brain's spatial recognition center just as well as it does for color. Thus, "[t]hough there are some brain areas that are more sensitive to color than others, color vision emerges through the combined activity of neurons in many different areas,"[9] say European researchers Karl Gegenfurtner and Daniel Kiper.

COLORFUL MEMORIES

"[C]olor is what the eye sees best," says Gegenfurtner. Of all the inputs the eye receives, small spots of red light are the most easily detected. Although the eye and brain may fail to discern subtle changes in illumination, even slight shifts on the red-green spectrum are obvious. Gegenfurtner has shown that color sig-

Different Types of Memories

Some memories are fleeting, while others last a lifetime:

- *Sensory* memory lasts only 300 to 500 milliseconds. For a twinkle in time, your brain retains a nearly perfect image of what you see (iconic memory) or hear (echoic memory).
- *Primary*, or *short-term*, memory lasts up to 30 seconds. The average person can store about seven bits of information in short-term memory. That lets you remember a phone number long enough to make a call.
- *Working memory* is the system that lets you keep track of a conversation, remember where you parked your car, or work through several steps in a math problem. Working memory lasts for minutes, hours, or days—as long as you need it—and then disappears.
- *Long-term* memory is more or less permanent. In a lifetime, a brain may retain one quadrillion separate bits of information. That's a one with 24 zeroes after it, or a million multiplied by itself four times. Memories may also be classified by their content:
  - *Verbal:* language
  - *Spatial:* the forms, positions, and movement of objects
  - *Episodic:* times, places, and events of our lives
  - *Declarative:* facts; for example, Lincoln wrote the Gettysburg Address
  - *Procedural:* how to do anything from hammering a nail to composing a sonnet
  - *Habit:* procedural memory associated with some reward
  - *Motor learning:* the process that improves movement skills through practice

nificantly improves the speed at which images can be recognized. It also improves the memory of those images. "How does the brain do it?" he asks.[10]

No one knows the answer to that question, but future research will undoubtedly probe not only the nature of color perception, but also the nature of memory itself. Ask a scientist what memory is, and you'll probably get long-term potentiation (LTP) as an answer. The idea goes something like this: When a nerve impulse reaches the end of an axon, it triggers the release of a neurotransmitter into the gap of the synapse. When the neurotransmitter attaches to the dendrite of the next neuron, it starts an impulse in the second cell. If this happens many times, the signal is strengthened, maybe permanently. In this way, neurons become conditioned to respond strongly to signals they have received many times before. Or, in simpler terms, "Neurons that fire together wire together."[11]

The process of long-term potentiation requires the chemical action of neurotransmitters. Among the hundreds of neurotransmitters that operate in the brain, memory synapses in the hippocampus and cortex make the most use of the neurotransmitter glutamate. It takes a lot of glutamate to trigger an impulse, but once a synapse is excited by glutamate several times, it becomes persistently more excitable. "The synapse has learned something; that is, it was 'potentiated' or strengthened. From then on, it takes less of a signal to recall a memory," explains neuroscientist Robert Sapolsky.[12]

Repeated signals increase the concentration of a messenger molecule in the receiving neuron. The messenger is cyclic adenosine monophosphate (cAMP). It relays a signal to the nucleus of the neuron and triggers a series of chemical reactions that produce the memory protein CREB. CREB, in turn, promotes the production of still other proteins that operate in the synaptic gap and strengthen a memory.[13] Yet more proteins may be responsible for maintaining a memory. New York researchers studied an enzyme called PKMz. When it's present, a memory persists. When its action is blocked, a memory is lost.[14]

While researchers generally agree on the LTP processes that form and maintain memories, they also recognize that not all memories are the same. The different types of memories (see sidebar) may explain, in part, why we remember naturally colored scenes better than black-and-white ones, even when the luminance is the same. Although their conclusions are controversial, the researchers at Max Planck who showed people the colored and gray-scale pictures think that the color advantage begins in the earliest stages of visual processing. They say color provides "sensory facilitation"; that is, we detect the shapes, sizes, and orientations of objects more easily and more quickly when we see them in color. The detection happens during the sensory or iconic phase of memory formation, within a mere 50 milliseconds. But iconic improvement doesn't explain all the memory advantage that color confers. The German researchers suggest that memories of objects are stored in an achromatic, black-and-white form in one memory system and in a surface-based, episodic form in another. However, memories interact, and color is no exception; the recognition of objects and scenes improves only if the colors are the usual, natural, expected ones.

So you'll remember a photograph of a yellow banana, but you'll probably forget a blue one. In the brain, "color is intricately linked to the identity of an object," the researchers say.[15] Color perception affects memory, and memory affects color perception. Our prior knowledge of the world changes the color we see.

Chapter 18

Sizing Things Up

I live by a mountain lake. It freezes in the winter. That's one of the best times to observe wildlife: foxes, eagles, white-tailed deer. I often spot the animals in the distance, far on the other side of the icy expanse. They look like tiny specks, but my brain recognizes them for what they are, and I know their true size. How? Our eyes tell us only part of what we need to know. The brain does the rest, making inferences about what's "out there."

Some of those inferences are all wrong.

THE MOON ILLUSION

When I was a child, I used to rush home from school to watch my favorite television show, *The Kate Smith Hour.* Her theme song was "When the Moon Comes Over the Mountain." My imagination pictured the moon rising over East River Mountain (where I lived) with every glissando that plus-sized crooner warbled. The image was clear in my mind. As it topped the summit, the moon looked huge, as big as the mountain itself. As it ascended into the sky, its size diminished. Miles above me, the moon shrank.

It wasn't until my adult years, however, that I heard about, and began to wonder about, the moon illusion. My problem was that I didn't realize it was an illusion. I always assumed—if I thought about it at all—that the moon moved farther away as it rose, appearing smaller as all objects do when their distance from the eye increases. I came to realize that such an explanation is nonsense. The moon moves no farther away when the earth turns. Its change in size is an illusion.

Since I've begun to investigate such matters, I found at least six plausible explanations for the moon illusion. Some of them require mathematics far beyond my grasp, but the one that makes intuitive sense to me involves visual clues. When the moon is on the horizon, it appears to be as far away as the horizon. Therefore, the brain "judges" its size as roughly mountain size. When the moon rises in the sky, the visual reference point is lost. The moon is a small sphere floating in a vastly larger empty field, so the brain "decides" the moon's size has decreased.

So, if that hypothesis is correct, the moon's size appears to change only because its position relative to the landscape changes. Landscape artists use this principle every time they draw perspective lines, and even primitive painters have the idea: Put an object toward the top of a canvas and the brain assumes it is farther away than an object at the bottom.

Are optical illusions such as the moon illusion anything but curiosities? Scott Murray at the University of Washington thinks they are more than tricks. He sees them as clues to how the visual-processing regions of the brain work, and he has used them to demonstrate some facts that existing theories about vision can't explain. One of them is what happens in the brain's visual centers when we "size things up."

I head off to Seattle to learn more.

A VISIT WITH SCOTT MURRAY

To enter the psychology department of the University of Washington in Seattle is to enter a rat's maze of identical corridors and doors. Guthrie Hall, where Murray and several dozen other psychologists work, is enormous, with a labyrinth of barren, dark hallways jutting off in this direction and that; the only distinctive architectural feature is a solid wall of vertically textured concrete that constitutes the outer perimeter. When I go there to meet Scott Murray, I wander for a while, wondering behind which wooden door my reward of cheese might lie.

I eventually find Murray and we ride a service elevator as featureless as the office and conference room where we settle down to talk. No posters adorn the institutional walls. No cushions soften the metal chairs. No flowers grace the utility table where we lean our elbows. Jumbles of scientific journals are strewn helter-skelter on warehouse-style shelves. In an environment as austere as this one, small signs of human habitation assume major significance. The conference room sports a Foosball table, and Murray has chanced a nod toward decorating his office with cow-print throw pillows and a framed photograph of his wife paddling an ocean kayak. We are, after all, only miles from the natural splendor of Puget Sound, so who needs artwork?

Murray is tall, rangy, bearded. He's a soft-spoken academician and the quintessential scientist. His love for learning shines, although he eschews both demonstrative gestures and verbal excess. "Personally, I could probably be interested in any kind of scientific topic," he says. "I just like science and thinking about the scientific process."[1]

Early in his studies, Murray struck an interest in what he calls intermediate-level vision problems: how we group items in a scene or how we recognize objects and faces. For a time, he studied hearing, which works in a hierarchical fashion much like vision. But the auditory region of the brain is very small and hard to measure with techniques such as functional magnetic resonance imaging (fMRI). The visual system has conveniently organized itself at a resolution that suits MRI investigative methods perfectly.

Vision research also attracted Murray because different people, working with different methods, can interrelate their findings, he says. Animal studies, perceptual analyses, cellular morphology, brain imaging—all can come together to create what he hopes will someday coalesce into a meaningful understanding of how vision works.

So how does it work? I ask

PROBING THE V'S

Do this. Run your hand up the back of your neck until you come to a bump. Place your palm on the bump, and you're squarely over your occipital lobe, the region of the brain that handles input from the eyes. The region is organized in a hierarchical fashion: first-level processing in the V1 (somewhere around your palm), higher-level processing in V2 farther out (think the base of your fingers), V3 farther out along your fingers, and so on. Vision researchers such as Murray spend a lot of time trying to figure out what's going on in V1, V2, V3—all the way out to V8.

The existing theory that everybody works with (see Chapter 16) suggests that V1 receives an exact copy of the neural stimulation that comes from the retina. That means (perhaps oversimplifying) that every spot of light that triggers a neural firing in the retina should correspond to a pinpointed neural firing in the V1. The V1, thus, should get an exact duplicate record of what the retina "sees." The V1 is expected to make that raw data available to the higher visual-processing centers, V2 and so on, where the meaning of the data can then be parsed. Thus, until Murray and his team published a landmark paper in 2006, most people assumed that V1 maintains the neural equivalent of a photographic image. If our brains are going to be fooled about the size of the moon or some other optical illusion, most experts thought, then the trickery was going on higher up in the V2, V3, or above.

Murray has challenged that view. He says he was explaining the idea of a "photographic" V1 in a speech he was writing one day. He wrote that the fundamental nature of V1 was known and moved on to more interesting topics, but his own words gave him pause. He described some size illusions and some color illusions and stated, "We know at the level of V1 that the two representations

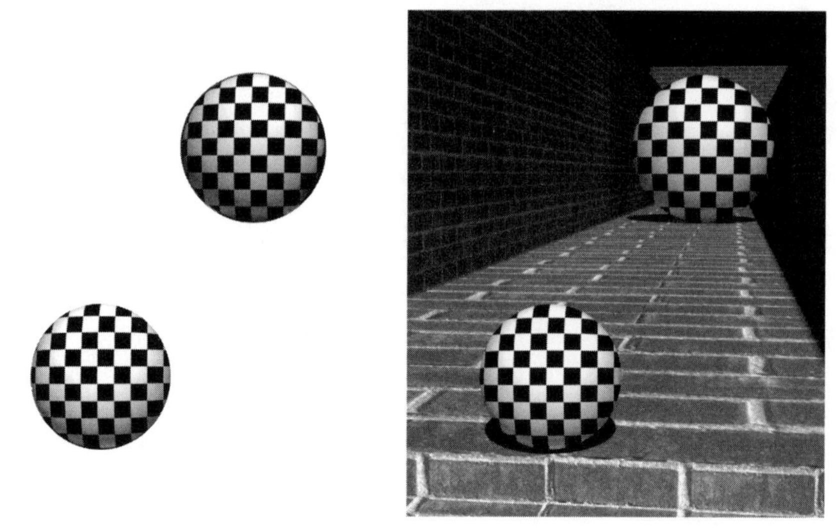

Figure 1

Credit: Huseyin Boyaci, Bilkent University, Turkey.

[will] be the same." But would they? "I had that phrase in my talk, but we didn't really know that they would be the same size in V1, so I thought I should go back and test that before I said it." That's what Murray did. His breakthrough experiment was supposed to be nothing more than a control experiment—one meant to verify what vision researchers already knew.

Here's what Murray did. He captured fMRI scans of the brains of human volunteers as they viewed the two images (among others) above (Figure 1). In the picture without a background, it's obvious that the two balls are the same size, and Murray's research subjects reported seeing them that way. Their self-reports were confirmed in scans of the V1. The two balls occupied the same amount of neural space—spots that the MRI detects as increased brain activity in a particular area. The images of the balls produced a "retinotopic image," an exact duplicate of the information sent from the retina to the V1.

Where conventional wisdom failed was in predicting the V1 map of the balls in the brick hallway image. What Murray expected was the same result as with the no-background balls. The V1 should have received and recorded what the retina "saw." But it did not. The volunteers reported that the ball in the back was at least 17 percent larger than the one in the front. The scans of their V1's showed that, too. The regions of activation were 17 percent larger, just as when the people viewed balls that actually were bigger. The illusion produced differences in V1 activity just as a real size differences did.[2]

What is revolutionary about Murray's results? No one expected size judgments to be made in (or even before) the V1. Sizing things up should be a higher-

What You See Is What You Imagine

Somewhere along the road to perception lies the stopover called imagination. In terms of brain activity, imagining a thing is not much different from actually seeing that thing. But can imagination actually change what we see?

New evidence from researchers at Vanderbilt University says yes. "Imagining something changes vision both while you are imagining it and later on," says psychologist Joel Pearson.[3] He has shown that "top-down" expectations, imaginings, or memories can change what the eyes see. "You might think you need to imagine something ten times or one hundred times before it has an impact," says Pearson's co-researcher Frank Tong. "Our results show that even a single instance of imagery can tilt how you see the world one way or another."[4]

In the Vanderbilt experiments, subjects imagined simple patterns of vertical or horizontal stripes. The imaginings stimulated the primary visual areas of the brain. Then the investigators showed their subjects green, horizontal stripes in one eye and red, vertical stripes in the other. Normally, such competing images cause viewers to perceive both patterns, with abrupt switches back and forth as the brain attends to data coming in from one eye and then the other. But that's not what happened here. The subjects saw the pattern they had imagined and missed the contradictory data coming into one eye.

The effect of imagination was about the same as when subjects actually saw a faint image of one of the patterns between trials. The effect was stronger if the subjects either viewed or imagined a pattern for a longer time. The Vanderbilt team says that both seeing and imagining generate a "perceptual trace" that modifies what is seen later. As expected, some people are more susceptible to the effects of imagination than are others.[5]

level process, but it isn't. The perceived size of an object—not its true size—determines the amount of processing space allocated to an image in the V1. "You don't have conscious access to the size of the image on the retina," Murray explains. "If you believe the hierarchical model of the visual system, this result shouldn't occur." But it does.

FEEDBACK

How might this result be explained? Murray says no one knows, but he has some ideas—ideas that came from some of his past research. In an earlier project, he asked volunteers to look at groups of stationary or moving dots or lines. Sometimes the dots and lines were random. At other times, they formed geometric two-dimensional or three-dimensional shapes. Murray used fMRI to see what his subjects' visual areas were doing when they viewed the irregular or patterned forms. In his analysis, he compared activity in the V1 with what was going on in the lateral occipital complex (LOC). The LOC lies just outside the V areas, but it is known to be active when shape recognition is occurring.

When the shapes grew more regular and geometric, activity in the LOC increased. That was expected. The surprise came when activity in the LOC and the V1 were compared. If the V1 were simply holding information from the retina and passing it on to other processing regions, then its activity should be constant, no matter what occurs elsewhere. That's not what Murray found. Instead, as LOC activity increased, V1 activity decreased. Murray concluded that the LOC was regulating the activity of the V1 via feedback. That is, neuronal projections from the LOC to the V1 were altering V1's action.[6] It's as if the LOC says, "V1, you don't have to work so hard. I recognize this as a shape, so I'll take over, and you can take a break." So, what Murray showed is that the V1 is "seeing" what the higher visual-processing regions "tell it to."

This feedback results from an actual structural connection between and among neurons. "A cell body in one visual area sends an axon that synapses back in V1. That is the feedback," Murray explains. The neuronal projections that carry the feedback message may go directly from the LOC to the V1. Or they may be indirect, traveling through other V areas before they reach V1. Murray says that animal experiments provide evidence for both direct and indirect connections. But however the feedback message travels, it gets back to the primary visual cortex, V1. "Activity in early visual areas is reduced as a result of grouping processes performed in higher areas," Murray says. "Inferences of high-level areas are subtracted from incoming sensory information in lower areas through cortical feedback."

BACK TO THE MOON ILLUSION

Murray's feedback explanation makes sense, but I wonder why I can't use a bit of conscious feedback to modify what's going on in my V1 when I misjudge the size of the moon. It seems that knowing that the moon is the same size at the horizon as it is high in the sky should allow me to override my visual perception. I should be able to coax my recalcitrant brain into accepting the truth. But I cannot, no matter how hard I try, see the moons as the same size, any more than I can coerce my eyes and brain into seeing Murray's two balls in the corridor as the same size. Why can't I force myself to see things as they really are? Why do these illusions persist no matter how hard we consciously attempt to correct them?

"The reason we can't make them go away," Murray explains, "is [that] the neural representations for these two stimuli are different at the very earliest stages of the visual system." The brain can't start operating on a point-for-point representation of reality because it never has one in the first place. The environment is interpreted long before the level of conscious awareness. The feedback that adapts us to the environment we expect—not the one that is really there—is hardwired into our visual system. "The perceptual system doesn't care about what is on the retina," Murray says. "What it cares about is what is out there in the world. It isn't trying to explain information on your retina. It's just trying to use the information on your retina to explain what is out there."

Chapter 19
On the Move

Those who lose a sense have much to teach us about it. Consider the cases of the neurological patients known only as R.A. and F.D.

R.A. was a retired computer manager who suffered a right-hemisphere stroke at the age of sixty-six. For three weeks after his stroke, his speech was slurred and his left arm and leg were weak. He complained of a severe headache at the back of his head, on the right side. His medical history was grim; he had heart disease, diabetes, high blood pressure, and high cholesterol. After his stroke, he entered a rehabilitation hospital in hopes of regaining strength on his left side. He did, but there was more to R.A.'s story. The stroke affected his motion vision. He complained that, when riding in a car, he could no longer judge the speed or direction of other vehicles traveling on the road.[1]

F.D. was a forty-one-year-old social worker. He suffered a mild stroke in his left hemisphere. He experienced weakness on his right side, but the weakness disappeared in a few days. He had trouble naming objects for a few weeks, but that difficulty also resolved itself. He could discriminate the speed and direction of movement with no difficulty. His perception of two-dimensional and three-dimensional objects was normal. His color vision was normal, too. Yet he complained of feeling disturbed by visually cluttered moving scenes, such as the countryside passing by when he was riding in a car.[2]

Both men recovered strength and speech, but their visual problems lingered —so much so that several years later, they contacted the Brain and Vision Research Laboratory and NeuroVisual Clinic at Boston University. "When people can't get anywhere, they end up with me," says Lucia Vaina, director of the laboratory and clinic. Vaina is a rare bird in academia: a neurologist with affiliations at Massachusetts General Hospital and Harvard Medical School who works in Boston University's Department of Biomedical Engineering.

"Some of my patients are blind to only certain types of motion," Vaina says, and that's why I've come to talk with her on this rainy day in May. She's invented and developed tests of motion vision, measured motion vision in normal people, and worked with patients like R.A. and F.D. to discover how the brain's visual system processes and perceives motion. I'm eager to learn about her work.

TYPES OF MOTION VISION

The first thing I ask Vaina is whether all motion vision is the same. It isn't. Translational motion means up, down, left, or right. Circular motion is a wheel going around. Radial motion is something coming toward you (it's expanding) or moving away (it's contracting). These distinctions are of more than academic interest. "When you move," Vaina explains, "the world seems to be expanding. It grows bigger as you approach things. If you are going backwards, it gets smaller. It is contracting, so that is the importance of expansion-contraction in navigation." Vaina has patients who cannot differentiate coming from going. One, who can't perceive motion at all, nevertheless gets around, using landmarks. Comparing the change in size and position of a landmark, the patient infers whether she has moved toward an object or away from it.

The most basic distinction among types of motion vision is first-order and second-order.

First-order motion is easy to define: it's motion perceived because of a change in luminance across time and space. Imagine a black bug crawling across white paper, and you have it. You perceive the bug's motion because the black and the white reflect different amounts of light to your eyes. When the luminance pattern changes from one time to the next and from one point on the paper to the next, you perceive first-order motion.

Second-order motion is a little more difficult to explain. It is defined not by a change in luminance but by a change in texture, flicker, or contrast. An example of second-order motion is a butterfly flying among the trees. The luminance of the butterfly and its spotted background is the same, but the difference in texture between the butterfly's wing tips and the background of leaves reveals the insect's motion.

To get the idea of how second-order motion vision works, imagine one hundred bugs. They are black on their backs and white on their bellies. The bugs line up single file on a square grid of ten columns and ten rows. In the beginning, some are black side up and some are white side up in no particular order. On a signal, all the bugs in column 1 flip over, so where there was black, there is now white and vice versa. The bugs flip so fast that you don't see the flip, only

The Amazing Case of Biological Motion

In 1973 Gunnar Johansson attached small lights to the head, shoulders, arm, and leg joints of an actor dressed all in black. He photographed the actor moving in total darkness. In the resultant film, no sign of the actor's body shape or size was present. Only the lights were visible, but observers recognized the moving human form immediately.[3] Thus, the term *biological motion* was born. It's a special kind of motion vision, unique unto itself. Minimal clues reveal movement of the human body, whether it's walking, running, hopping, or jumping.[4] Even the gender of the moving human is easily identified by the motion-vision system. Male figures appears to face the viewer. Female figures seem to be moving away.[5] Some of Vaina's patients who are seriously impaired in simpler types of motion vision can nonetheless identify the human body in motion, using as clues only a few spots of light or dark on a computer screen.

Now there's evidence that the human brain's propensity for recognizing other humans in motion is inborn. Researchers in Italy tested two-day-old babies with light displays of biological and nonbiological motion. The infants could tell the difference between the two, and they looked longer at the biological motion. Their preference was orientation dependent; the babies gazed longer at upright displays of biological motion than at upside down ones.[6]

the color change. Now, within milliseconds, the bugs in column 2 flip black to white and white to black, and then column 3, and then column 4, and so on, all the way to column 10. Then they reverse the process and start flipping again, starting with column 10 and moving back to column 1. While watching this, you would perceive not simply a reversal of the colors, but a vertical line moving from left to right and then right to left. There would, in reality, be no moving line, but you'd see one, simply because the contrast between the successive columns was changing.[7]

TESTING MOTION VISION

Double dissociation. That's what her research on motion vision is all about, Vaina tells me. Double dissociation "means I have two tasks: task A and task B," she says. "You can do A but not B, or you can do B but not A. That means that in the brain, A and B are not necessarily connected. You can dissociate them.

You can doubly dissociate them: A but not B, B but not A. That tells us a lot about the organization of the motion subsystem."

Different types of motion vision are handled in different parts of the visual cortex, and from patients such as R.A. and F.D., Vaina has learned that those types can be doubly dissociated. A brain injury such as a stroke can rob a person of one type of motion vision but not another. Furthermore, the lost of a "simple" type of motion vision, such as translational motion, does not necessarily mean loss of a "complex" form, such as radial. In fact, the opposite can occur; "simple" motion vision can be lost, while the ability to perform "difficult" tasks remains normal. That means that motion vision is not strictly hierarchical, Vaina explains. Information need not pass from the lower visual centers upward, in a single direction. Instead, the perception of complex motions can be managed independently of other motion-detection tasks. A but not B; B but not A.

How does Vaina look for these double dissociations? She has invented and perfected a series of visual tasks that she uses to test individuals—both normal and neurologically impaired—on different types of motion vision. Her tests, which were not possible until computers became as common as lead pencils, present a test subject with a stimulus on a screen. The stimulus is tightly and narrowly defined, so only a single element of motion vision is assessed in any one test.

An example of one such test is coherence. Imagine one hundred bees in a jar. Most of them move about randomly, but a certain number fly in a single direction—perhaps left or right, up or down. How many of them have to be traveling in a same direction for the viewer to perceive the direction of their motion? For people with normal motion vision, the answer lies somewhere around 10 percent. Only ten of those hundred bees need to be flying toward the right for the test subject to say, "Right." But for patients with brain lesions in certain regions of the occipital lobe, where visual processing occurs, that number may rise to 30 or 40 percent. Yet such patients, highly impaired on coherence, may score in the normal range for other motion-vision tests, even more complex ones.

Vaina uses a host of such tests to assess the motion vision of her subjects, looking at but a single element of motion vision at a time. She can test for (among others) sensitivity to contrast, detection of shape, wavelength discrimination, color naming, binocular vision, location of a point, orientation of a line, speed discrimination, inference of three-dimensional structure from motion, face recognition, matching silhouettes to objects, object recognition, object memory, topographic memory, and biological motion (see sidebar). Her patients can vary in their capacities on all those tests, showing normal performance in one and impaired performance in another, in a doubly dissociated fashion.

Patient R.A. proved to be impaired in first-order motion vision, but not second-order. Vaina worked with him for two years after his stroke. His depth perception was normal, but he had problems recognizing objects. Initially, his

See the Motion That Isn't There?

Two regions of the brain specialize in motion vision: MT (medial temporal cortex, also called V5) and MST (medial superior temporal cortex). Hundreds of brain-imaging studies have shown increased activity in MT and MST when people view objects in motion, or even when they *imagine* a moving object.

What happens if motion is not shown, but implied? Researchers at the Massachusetts Institute of Technology captured fMRI images of brain activity while volunteers viewed still photographs of athletes stopped mid-motion, stationary athletes, people at rest, houses, and landscapes. Any human image evoked greater activation in MT and MST than did the houses or landscapes, but the greatest activity came when subjects saw athletes stopped at some point in a continuous movement, such as throwing a shot put.

The researchers explained this as a "top-down" effect. They suggested that the inference of motion is not constructed in MT/MST directly; instead, other, "higher" areas of the brain categorize the object, link it to learned associations about the object, and subsequently moderate MT/MST activity to perceive motion that isn't there. This top-down association happens automatically, with no conscious awareness, attention, or deliberation on the part of the viewer.[8]

Further evidence for this idea came in 2007 when scientists at the Salk Institute in La Jolla, California, trained monkeys to associate a stationary arrow with a display of moving dots. Arrows mean nothing to monkeys, but just to be safe, some of the monkeys were trained to associate an arrow pointing up with dots moving down and vice versa. Later, when signals from the animals' neurons were recorded electronically, the sight of the static arrow produced activity in MT and MST, exactly as the moving dots had. The association comes from learning, the researchers say. Other regions of the brain "tell" the MT to "see" motion that isn't there.[9]

right hemisphere lesion impaired his vision on the left side of his body, but he soon recovered that capacity. However, his first-order motion remained deficient. He could detect movement from changes in texture, flicker, or contrast, but changes in luminance defied him.

Patient F.D. showed the opposite pattern. His first-order motion vision remained normal after his stroke, but his second-order was impaired. Vaina tested him with stimuli such as flickering bars. He could not see their movement. She tried rotating patterns of dots moving clockwise and counterclockwise. F.D. could not discern their motion. For the most part, he was able to navigate normally, because he found ways of compensating for his loss of second-order motion vision. He was, however, sometimes aware of his second-order deficit. When his son's soccer ball rolled into the garden, he said, he could not trace its path. The luminance of the rolling black-and-white shapes was the same as that of the dappled light and shadows of trees on grass. Unable to use contrast, texture, and flicker cues, and robbed of luminance cues, F.D. could not see the ball.

STEAL FROM YOUR NEIGHBOR

R.A. and F.D. had lesions in different areas of their brains. Comparing R.A. and F.D., along with brain images from dozens of other patients, has enabled Vaina to identify areas of the brain that specialize in the different types of motion vision. Area V2, where R.A. experienced his lesion, processes first-order stimuli, and its projections send information to the higher visual-processing centers, notably MT (V5).[10] Vaina believes that second-order motion may be localized in a region in the visual cortex that lies next to, and slightly above, MT. The region does not require direct inputs from V1 and V3. The fact that MT responds to any

Blindsight

After a stroke, the eyes continue working normally, but if the stroke killed neurons in the brain's visual centers, then vision may be impaired, even destroyed. Nevertheless, the eyes are still sending visual information to the brain—a condition known as blindsight. Can undamaged parts of the brain be retrained to take over for the parts that were lost? Until recently, few treatment centers had tried the intensive rehabilitation for vision that is routine for the loss of mobility or speech in stroke patients. Then, in 2009, researchers at the University of Rochester Eye Institute demonstrated what vigorous visual exercises could achieve. The scientists worked with five patients who had gone partially blind after strokes. All five had damage in V1, often considered the "gateway" through which visual information must enter the brain. The patients worked daily on a series of visual exercises presented on a computer screen. One of the exercises "forced" the brain to "guess" when dots were moving in an area of the visual field that the patient couldn't consciously see. After nine months or more of training, all the patients regained some vision, some of them even returning to such prestroke activities as shopping and driving.[11]

sort of motion, either first- or second-order, may explain why R.A. could perform second-order tasks, although he was impaired on first-order ones.[12]

Vaina says her research has two goals: (1) to better understand motion vision in the brain, and (2) to apply that understanding to the rehabilitation of patients such as R.A. and F.D. In her clinic, some of the vision tests Vaina has devised do double duty as training tools. Patients come in to work at her computers, attempting to improve on those specific tasks on which they consistently fall short. Many of her patients do improve over time, and their training gains translate into better day-to-day functioning. One patient, Vaina says, was able to resume driving again, once his training on rotational motion vision allowed him to discriminate the movement of vehicles on Boston's many roundabouts.

What's more, Vaina's latest research shows precisely how the brain may reorganize itself after a stroke or other injury to compensate for capacities that have been lost. Neural plasticity, Vaina explains, is a localized process. The brain's motion-processing regions, once damaged, can't appropriate neural territory in a distant region of the brain, but they can "steal from their neighbors," diverting neurons that previously performed one task to perform another. She shows me a PowerPoint presentation of unpublished research; it will become a scientific paper she hasn't gotten around to writing yet. It is a report on a patient identified as J.S., who suffered a right-hemisphere stroke in the visual-processing areas. Vaina worked with J.S. and captured MRI images of his occipital lobe at three, eight, and eleven months after the stroke. Using some complex mathematics, she measured changes in the size of the visual-processing areas, specifically those designated V2d, V3, VP, V3a, and V4. In J.S.'s left hemisphere, where the stroke did no damage, the size of those regions changed little over time. However, on the brain's right side, areas VP and V4 shifted slightly, while V2d, V3, and V3a grew dramatically—by as much as almost 7 millimeters. The growth occurred because the visual regions impinged on their neighbors. Predictably, as those visual areas expanded, J.S.'s performance on motion vision tasks also improved.

The same thing must have happened in the brains of R.A. and F.D. As they recovered from their strokes, various brain regions "stole from their neighbors," and neurons reorganized themselves to regain some, if not all, of what was lost. "It is hard for me to acknowledge that this really works," Vaina says. "I believe scientifically that it works . . . but to say to my patients, 'It is going to work,' I can't do that. It is ethics. This may not work. If it does work, so much the better. If not, that is that."

Still, whether rehabilitative applications succeed or fail, each of Vaina's patients advances our understanding of motion vision. And so does Vaina herself. "My business is to find out how you do what you do. . . . That allows me to look into neural rehabilitation and neural plasticity in order to help people who no longer can do things," she says.

Chapter 20

Vision and the Video Game

I recently conducted a nonscientific email survey on a nonrandom, nonrepresentative sample of friends and family members. I posed this problem: Suppose you are considering laparoscopic surgery. Only two surgeons perform such surgery in your town, and you have no reason to doubt the competence of either. Both surgeons are male and the same age. The only difference between them is that one is an avid video game player in his free time. The other is not. Which surgeon would you choose? Would your choice change if one had more experience than the other did?

My questions evoked a flurry of Internet exchanges. Some respondents refused to play the game. They insisted on investigating other variables they considered more relevant, such as surgical success records or infection rates. Some thought experience was more important than game playing: "I would choose the doctor with the most experience, so in case of an unexpected problem, he would know what to do," one responder said.

A psychotherapist friend admitted that she sees game players only as troubled young people in counseling. "Most of the video gamers I meet," she said, "feel alienated from people; they construct an alternative reality for themselves. The range of feelings that would normally be present in human interaction is absent from their artificial world." Presumably, she wouldn't want such a person cutting into her body.

Some respondents rejected the game player—experienced or not—for personality reasons. "Gamers are dorks," commented one relative. Others doubted the game player's work ethic: "What is this free time he has so much of? Why isn't he outside mowing the lawn?"

But some respondents—surprisingly neither the youngest nor the trendiest—chose the game player hands-down. "The video game player would be my choice because of the quick, nimble fingers. Manual dexterity in a surgeon would be a plus. Also the ability to make quick decisions and react accordingly would be useful," said one friend. "I would probably take the game player, even if he had less experience," said another. "He may be better suited to do the surgery via the kinds of images video games demand."

Aside from the diversity of responses, what interested me the most was the acceptance of video game playing as a relevant variable. While some people said they'd consider other factors as more important, no one thought game playing didn't matter. Implicit in the answers was the assumption that game playing changes the brain—and the person who owns it. The only dispute is how. That's where brain research comes in.

EXPERIMENTS ON GAMERS

One of today's young and innovative brain researchers is cognitive psychologist Shawn Green. When he entered vision research, he started with what he knew: the video game. Himself an avid game player and a lover of active sports of all kinds, Green began as an undergraduate at the University of Rochester to investigate how video games affect the visual skill of expert players and of beginners who learn to play the games. His graduate-level research resulted in papers in prestigious journals such as *Nature* and *Cognition*. In 2008, at age twenty-seven, Green became a postdoctoral research associate at the University of Minnesota.

I go to visit him at his office one bright spring morning. I can see the Mississippi River from his window. He is waiting for me, tossing a football from one hand to the other while simultaneously answering emails from the Brain and Vision Lab back at the University of Rochester, where he still has experiments running. He is tall and lodgepole thin, with baby fine brown hair curling on his shoulders, sunglasses pushed to the top of his head, and a functioning pair (untinted) perched on his nose. His face is ethereal; he reminds me of an angel in a Renaissance painting. I congratulate him on his publications. "Writing the papers is my least favorite part of science," he replies. "I'd be happy just knowing the answers myself."

But Green shows no reluctance in sharing his answers with me. Visual attention interests him most, he explains, but he doesn't define attention in the ordinary way. It's not purposely "paying attention," the way a teacher might admonish a student to do. Instead, it's the focusing on an object, event, or feature within the visual field that happens automatically in the brain. The trick for the researcher lies in using a visual test that is specific enough to measure a single, reproducible aspect of visual attention. I ask Green to walk me through some of the tests he used on video game players, and he does:

1. *The Attentional Blink.* Imagine you are looking at a screen. One by one, letters appear, then wink out. Most of the letters are black, but one is white. Some time after the white one, an X may or may not appear. Your job is to identify the white letter, then tell whether the X appeared. That seems simple enough, but test results show some anomalous patterns. If the X comes immediately after the white letter, you have a better than average chance of seeing it; the brain tends to chunk the X in with the white letter, a phenomenon called "lag one sparing." But if the X appears within 200 to 500 milliseconds after the white letter, many people miss it. For an instant after one stimulus is perceived, the visual system is "blind" to another. That's the attentional blink.

    Green says a number of hypotheses might explain the attentional blink, but he offers a simple one. "You are switching to a new task. You are going from an identification task to a detection task, so you need a little time." But some people do better overall, and some people need less time than others do. Video game players outperform nongame players on this test. In the lag one part of the test, where the X comes immediately, they outperform nongame players by better than 30 percent. In the attentional blink window, they do better by some 40 percent. As more time passes between the white letter and the X, the nongame players catch up, but they never did better than the game players.[1] "I don't have an attentional blink," says Green.
2. *Enumeration.* Now back to imagining that screen. This time, squares flash on the screen for 50 milliseconds. As few as one or as many as twelve may appear. Your job is to say how many. Up to three, everyone tests pretty well. There are few errors. Above that, frequent game players differ markedly from nongame players. They score perfectly up to five, while nongame players are making lots of mistakes at that number. At six, game players start to make some mistakes, but they still score 10 to 20 percent better than nongame players all the way up to twelve.

    The smaller numbers are called the subitizing range. That's the range of numbers that the visual system can perceive without counting. Game players have a greater subitizing range than nongame players (five versus three). Above that, the brain switches to counting, and again game players excel. They count more accurately and make fewer mistakes.[2]
3. *Spatial Resolution.* Back to the viewing screen. This time you see three T's in a vertical row. Sometimes they are crowded close together. Sometimes they are farther apart. Your job is to tell whether the center T is right side up or upside down. When the top and bottom T's are far away from the center

T, you don't notice them, but as they begin getting closer, they start to interfere with your ability to discern the orientation of the central T. But not so for game players. They do better at every distance and with every size T. They even see the T better when it is presented alone, even though their corrected visual acuity (what glasses correct for) is the same as everyone else's.[3] Green thinks that timing may explain that effect. The T's appear on the screen for only 100 milliseconds.

4. *Multiple Object Tracking.* Return to the imaginary screen. This time you see sixteen circles moving randomly and bumping into each other. In the beginning, a small number of them are red and the rest are green, but in an instant, the red ones turn green. Your job is to keep track of the ones that were red. For one or two cued (red) ones, everyone does equally well, but game players start to outperform nongame players at three, and their advantage increases at four, five, and six.[4] "Once you get to seven, no one can do it anymore," Green says.
5. *Distracters.* You're back at that screen once more. This time, you are looking for either a diamond or a square that will appear at some point on a circle, placed like six number points on an analog clock. Sometimes other shapes appear on the clock points, making your diamond or your square harder to find. Sometimes a distracter shape appears inside or outside the circle. The distracter may be a diamond or a square, but the experimenter tells you to ignore the distracters. You must simply say "diamond" or "square"—whichever one appears at one of the six clock points.

    Despite those instructions, the brain pays attention to distracters. If the distracter shape is the same as the shape to be identified (square–square or diamond–diamond), respondents tend to answer more quickly. The distracter lessens their response time. Opposite distracters (square–diamond, diamond–square) lengthen response time. They slow the brain's visual system.

    The same is true of task difficulty—for most people, anyway. The greater the number of different shapes on the clock points, the more time the brain needs to find the square or the diamond. When tasks are very difficult, the distracter advantage disappears—again for most people. The brain's visual system is so busy looking for the square or diamond among all the other shapes, the "helpful" distracter no longer speeds their response time.

    But expert video game players don't perform that way on either measure. The difficult tasks don't faze them, and the same-shape distracter speeds them up, even when all the nonstimulus clock points are filled with wrong cues.[5] "Game players can process the outside things equally well regardless of how much extra stuff you put in there," Green says.

Two Sides of the Visual Coin

The distracter experiment poses an obvious question. Do game players have greater attentional capacity, or are they simply more easily distracted? Green thinks both are possible. Continuing to be influenced by distracters, even when the task is difficult, isn't necessarily an adaptive thing, he says. Game players may not filter what's relevant from what isn't as well as nongame players. Green explains attentional bottlenecks. "So if you've got this whole world out there, only some of it can get in. . . . It seems that video game players let more through." That may not always be a good thing.

However, Green is not convinced that game players are more distractible than are nongame players. He describes another experiment, one in which game players and nongame players respond to "go/no go" signals. When certain signals appear on a screen, subjects must either press a button ("go") or *not* press a button ("no go"). In one block of trials, the experimenters rig the test to see if their subjects are overly impulsive. "Go" signals appear repeatedly, so the viewer gets accustomed to pressing the button. Suddenly, a "no go" signal pops up. Impulsive people have trouble stopping themselves from pressing the button, but game players show no signs of greater impulsivity. They can block the "go" response when they need to.

The opposite test measures whether subjects' minds tend to wander from the task during periods of inactivity. Repeated "no go" cues appear, with only an occasional "go" signal thrown in. Game players are no more likely than anyone else to "blank out." Even when the task habituates them to inaction, game players can respond actively when they need to.[6]

"If game players have problems with staying on task or too much compulsiveness, they would show differences on this task, and they don't," Green says, "except they do the task a lot faster, without sacrificing accuracy."

YOUR SURGEON AND THE VIDEO GAME

So back to my bogus survey. Do video game players make better surgeons? Green suspects they do, and some evidence supports his belief. In 2007 researchers at Beth Israel Medical Center in New York assessed the laparoscopic skills of thirty-three surgical residents and attending physicians. The investigators compared the surgeons' skills on a training simulator, controlling mathematically for varying levels of surgical training and experience. They found that those who had played video games in the past for more than three hours per week made 37 percent fewer errors and finished the surgery 27 percent faster. The most

Serendipity in Science

Horace Walpole wrote in 1754: "I once read a silly fairy tale, called *The Three Princes of Serendip*: as their highnesses traveled, they were always making discoveries, by accidents and sagacity, of things they were not in quest of."[7] So the word *serendipity* was coined, and ever since then, the role of luck in science has been hotly debated. Some philosophers of science say no discovery is ever made without it. Others say no discovery is luck; new knowledge is invariably the product of a trained mind observing reality in a disciplined way.

Shawn Green's start as a vision researcher provides evidence that both arguments may have merit. As an undergraduate, he worked in Daphne Bavelier's lab at the University of Rochester. There he was charged with setting up an experiment called "useful field of view." The task requires the subject to fixate on a center point, then locate a rapidly flashed circle that can appear on any of eight spokes that radiate from the center point. The test is good at predicting accidents among elderly drivers. It tests a driver's ability to filter out distractions and detect hazards on the periphery.

"So I'm running this task on myself, and my results were off the scale. I was way too good at this task, so . . . I get my roommate in and he can do it perfectly, and I think I've got something wrong." Green brings in still another friend to try the task. This friend did rather poorly, "exactly the way the literature said he should." Green brought still other friends in to try the test and found nothing wrong with his programming. The difference lay in the subjects. Those who scored near perfect on the test were video game players. The low scorers were not.

"It became my senior thesis," Green says.

skilled game players made 47 percent fewer errors, performed 39 percent faster, and scored 41 percent better overall. Statistical tests showed that video game skill and past video game experience were significant predictors of laparoscopic skills. "Video games may be a practical teaching tool to help train surgeons," the researchers concluded.[8]

Green's work shows that training does indeed make a difference. Every summer, he runs training programs in which nongame players come to his lab in Rochester and participate in as many as fifty hours of training on a video game. Some volunteers learn an action video game such as Medal of Honor or Allied Assault. Others learn a more passive game such as Tetris (where rows of colored dominoes must be aligned) or SIMS 2 (in which each player lives the life of a mythical character in an imagined world).

On every measure of visual attention that Green has studied, people trained on the action video games improve. The control-group Tetris and SIMS 2 players don't. It's the action that makes the difference, Green thinks, although "off the shelf" video games have so many components, it's impossible to discern which features change the brain's visual-processing system. Green hopes to rectify that in his future research, designing and testing simple video games that develop a single aspect of visual processing.

"Right now, we don't know what in the brain is changing, but we have a lot of good ideas," he says. "Odds are, one of them is right."

PART FIVE
HEARING

All the sounds of the earth are like music.

—Oscar Hammerstein II
"Oh, What a Beautiful Mornin'"
(from *Oklahoma*)

Chapter 21

Hearing, Left and Right

With all this emphasis on the brain as the processor of sensory information, it's tempting to think of the sense organs themselves as passive receptors—the eyes little more than transmitters for wavelengths of light, the nose a mere collector of molecules. That view is erroneous: the sense organs are the first processors of sensory input. The image on the retina is formed through a series of chemical reactions, long before any impulse is ever dispatched toward the visual cortex. If the nose isn't working, neither can the brain's olfactory centers. The pattern of impulses generated in the nose is all the brain has to work with in sensing and identifying odors. Damage the taste buds in the tongue, and the sense of taste is lost. Damage the nerve endings or mechanoreceptors in the skin, and any of the various sensations of touch is gone. The sense organs do more than receive and transmit; they shape what the brain is and what it becomes. Without stimulation from the sensory nerves, specialized regions in the newborn brain never develop their normal functions.

Likewise, it's tempting to view the ear as a passive collector of sound vibrations. It looks like one, with its convoluted external shape reminiscent of the old-fashioned trumpets used as the first hearing aids and the flower-shaped horn of the gramophone. And so neuroscientists viewed it for a long time, doing little more than charting the course of vibratory pulses through the structures of the ear before getting to the interesting part—somewhere topside from the auditory nerve. Yet, if recent research gains support, the ears may be starting the processing of sound information before a single signal squeals off toward the brain. What's more, the right and left ears may do different kinds of processing, even beyond locating a sound's directional source. They may supply the kinds of differentiated inputs the two side of the brain need to develop normally. The left and right ears may actually induce the brain's left–right specialization for sounds by the kinds of stimuli they provide.

But that's getting ahead of the story. Before looking at the specializations of left and right ears, along with left and right brains, a summary of the basic processes of hearing is in order.

THE AUDITORY SYSTEM

Most schoolchildren learn the parts of the ear: the external auricles or pinnae, the chambers of the middle ear, and the sound-transmitting structures of the inner ear, along with the organs of balance it also houses.

Audible sound waves (which for humans lie between the low notes at 20 cycles per second and the high notes at 20,000) are herded into the auditory canal by the horn shape of the pinnae. Inside the canal, these waves of greater and lesser air pressure travel about three centimeters (a little more than an inch) before reaching the eardrum, the membranous surface and entryway to the middle ear. When sound waves set the eardrum to vibrating, they've already been modified by their interactions with the cartilaginous pinnae and the semiflexible canal.

Attached to the eardrum is a tiny bone, called the hammer because of its shape. The hammer attaches to another tiny bone, the anvil, which hooks up with yet another, the stirrup. These bones vibrate, each in its turn, and their vibration amplifies sound input, actually making sounds louder. The stirrup vibrates the oval window, the gateway to the inner ear.

Fluid fills the spiral-shaped cochlea of the inner ear. The fluid jiggles and bounces like ripples in a tide pool in response to the vibration of the oval window. Membranes lie between the three chambers within the cochlea. Attached to one of them—the basilar membrane—lie three to five rows (depending on how you count them) of tiny hair cells—some 16,000 to 20,000 of them in all. Those cells, the basilar membrane, and the cells that support them form the organ of Corti, sometimes called the microphone of the ear.

When the fluid in the organ of Corti vibrates, it vibrates the basilar membrane, and the hair cells on it wave about like seaweed in the surf. The waving hair cells stretch and contract. Their "dancing" helps amplify sounds, but that's not the only way these structures "pump up the volume." Somewhere between fifty and three hundred hairlike tubes called stereocilia project from the top of each hair cell. The stereocilia dance around, too, and their motion converts the electric signal generated by incoming sound into mechanical work, adding power to a sound and making it louder. The longer stereocilia respond best to low-pitched sounds. The short ones are best at amplifying high-pitched sounds.[1]

The hair cells are firmly attached at one end, but at the other end they float freely in fluid. At the floating end of each hair cell lies a puff of tiny filaments only three to five micrometers (millionths of a meter) long. Those filaments can touch another membrane, the tectorial (from the Latin for *roof*) membrane. As they move, some of the puffs at the top of the hair cells brush the tectorial membrane. This creates a shearing force that causes the hair cells to release a neurotransmitter. The neurotransmitter triggers a nerve impulse in neurons whose cell bodies make up the spiral ganglion (a ganglion is a group of neurons) that lies just outside the cochlea. The axons from the spiral ganglion form the auditory nerve (cranial nerve VIII). It carries signals from the ears to the cochlear nucleus in the brainstem.

In the cochlear nucleus, two paths diverge. The neurons toward the front of the structure respond to individual sounds and transmit incoming signals more or less unchanged; the rear part of the cochlear nucleus processes patterns, such as changes in pitch or rhythm.[2] From the cochlear nucleus, signals move upward to the superior olivary nucleus of the brainstem and the inferior colliculus in the midbrain. There appropriate responses to sound, such as turning the eyes, head, or body toward a sound, are generated. There also about half the nerve fibers cross to the opposite side of the brain, while the other half continue to carry the impulse on the same side. This crossover means that each ear sends signals to both sides of the brain.

The next way station is the medial geniculate nucleus of the thalamus (not far from the first stop for visual information, the lateral geniculate nucleus). The thalamus is the "gateway" to the cerebral cortex (for all the senses except smell, which has a direct line). The "gates" in the thalamus either let the signal through or they block it. The gating action of the thalamus may explain why you can focus attention on a single sound to the exclusion of others, such as listening to your conversant while ignoring the cacophony of a cocktail party. If allowed through, the sound impulse continues to the left and the right auditory cortices, which lie in the temporal lobes, just above the ears. In the cortices, the conscious recognition of a sound presumably occurs, as does the willful decision to act on it—or not.

What differentiates one sound from another? The answer begins back in the ear with the structure of the basilar membrane. It's stiff near the oval window, but increasingly flexible along its length. It gets wider and heavier as distance from the oval window increases, too. Those variations cause different regions of the membrane to vibrate in response to different frequencies. High-pitch (high-frequency) sounds cause the area near the oval window to vibrate most. Lower frequencies travel farther along the membrane and vibrate its more distant end. Thus, the "neural code" for pitch is the region of hair cells that generates an impulse, but other factors are involved. One is the rate of stimulation. High frequencies stimulate hair cells more rapidly than do low frequencies. Especially for the lower tones—up to about 3,000 cycles per second—the speed of impulse generation is important to the brain's interpretation of pitches. For higher frequencies, the place information from the basilar membrane seems to be enough to get the message through to the brain. In short, the distinction between middle C and high C is made in the ear, long before the impulse ever arrives in the auditory cortex to signal conscious knowledge of the difference.

Loudness is processed in a similar fashion, but it's a matter of amplitude: how far up and down the basilar membrane moves. The louder the sound, the greater the membrane's vertical oscillation. The greater amplitude increases the number of hair cells that make contact with the tectorial membrane. It also increases the speed at which impulses are generated.

Although no one has figured out exactly how the brain's sound-processing system is organized, neuroscientists suspect it may share some common features with the visual system, perhaps using one stream of impulses for locating a sound in space and another for identifying a sound. There may even be a third for handling the directions of moving sounds.[3] Logic suggests that circuits in the brain must allow auditory processing to communicate with memory systems, so that a sound can be consciously recognized and named. So, too, communication must flow back and forth between the visual system and the auditory system, so that we can, for example, relate the lip movements of speakers to the speech sounds they make or associate the roar of a jet engine with a 767 warming up on the tarmac. Similarly, complex connections must support the brain's ability to localize the source of a sound, tell two similar sounds from each other, gauge the timing and duration of sounds in an orderly sequence, differentiate and separate two or more competing sounds, and mentally filter out "noise" to home in on a single—comparatively faint—sound stimulus.

It's also obvious that information has to travel in both directions. Sound processing can't be a one-way street. The ears send sound signals to the brain, it's true, but the brain has to modulate what the ears are up to. If it didn't, the brain would be bombarded by a confusing cacophony of sounds from the environment, so attention to a single sound, or ignoring irrelevant sounds altogether, would be impossible. So, another road for traveling nerve impulses runs in the opposite direction from the auditory cortex. Termed the "descending pathway," it courses through all the stops made by the "ascending pathway," with tracks that cross from one side to the other, preserving the ability of both sides of the brain to communicate with both ears. The descending pathway ends in a distinctive type of nerve fibers that lie beneath the base of the hair cells in the cochlea. These fibers don't respond to sound stimuli. Quite the opposite. Primarily through the action of the neurotransmitter acetylcholine, they act as inhibitors, letting the hair cells know "when enough is enough . . . or too much." They actively block some impulses while allowing others to pass.

MUSIC TO MY (LEFT) EAR?

At one time it was commonly believed that the left brain handled language and the right processed music. That old saw has proved overly simplistic, as researchers have learned more about how the brain handles sound. While it's true that some of the brain's language centers lie in the left hemisphere in most people, other characteristics of sounds, such as pitch and volume, get their fair share of attention in both hemispheres, and not just in the auditory cortex of the temporal lobe. Regions of the frontal and parietal lobes are involved, too, in a sort of radiating pattern of cortical activation that moves outward as harmony, melody, and rhythm are perceived. Forward, behind, and to the sides of these sound-processing areas lie "associative regions." They "put together" the experience of a sound, whether it's speech, music, or an environmental sound.[4]

Despite this diffuse processing, specializations have been proved. The left brain excels in measuring the intervals between sounds and their duration; it's more proficient than the right in decoding the rapidly changing sounds of speech. The right takes center stage in pitch, musical notes, and melodies; it's a "big picture" thinker, zeroing in on music's melodic contours and patterns.[5]

If the brain's left and right hemispheres specialize, do the left and right ears specialize, too? Until recently, the answer was thought to be no, although behavioral studies have shown that we react faster and identify speech-type sounds more easily when they are played to the right ear; pure-tone pitches are more quickly and easily recognized when played to the left ear.[6] The explanation for such differences was once thought to lie in the specialized processing of the brain hemispheres, not in any difference in the early processing that occurs in the ears.

That changed in 2004 when Yvonne Sininger, of the Brain Research Institute at the University of California at Los Angeles, and Barbara Cone-Wesson, of the University of Arizona, published the results of their studies on infants. The researchers spent six years testing the hearing of nearly 1,600 newborns within days after birth. The scientists tested the babies' ears using a tiny probe that measured minute amplifications of sound in different parts of the ear. Sininger and Cone-Wesson played two kinds of sounds to the infants: sustained tones and a series of fast clicks timed to mimic speech. Measurements of sound volume in the inner ear showed that the cochlea amplified the clicks more in the infants' right ears than in their left. Conversely, the infants' ears "pumped up the volume" of sustained tones more in the left ear than in the right.

Sininger and Cone-Wesson concluded that the left and right ears were preferentially amplifying different sorts of sounds, showing an innate (because these babies were only days old) preference for the kind of sound input "preferred" by the hemisphere opposite each ear. Left–right auditory processing, they asserted, begins not in the brain, but in the ears.[7] If they are right, the reason may lie in those feedback pathways that run from the brainstem down to the ears. Some of them are known to be well developed at birth, so the selective ability of sound-sensitive neurons may be at least partially in place within the first hours of life.

Sininger and Cone-Wesson have demonstrated such an inhibitory feedback mechanism, suggesting that the right medial olivocochlear system, one of those switching stations in the brainstem, suppresses certain kinds of auditory responses in the left ear.[8] If that's true, then the brain is acting on the ears, but the ears are also acting on the brain. Differentiated inputs from the ears may direct how the right and left auditory cortices develop during childhood.

"Our findings demonstrate that auditory processing starts in the ear before it is ever seen in the brain," says Cone-Wesson. "Even at birth, the ear is structured to distinguish between different types of sound and to send [them] to the right place in the brain."[9]

Where Is That Sound Coming From?

Relatives. You gotta love 'em. Whenever a certain relative of mine (who shall remain unidentified, in the interest of family harmony) comes to visit, she uses the cordless telephone that normally rests on its charger in the kitchen. She carries it upstairs to the guest bedroom. She carries it downstairs to the TV room. She carries it out onto the back porch, happily chatting away all the while. The trouble comes later when the phone rings. We can all hear it, but where is the phone? Hard to tell. Maybe she left it in the living room. Maybe it's on the deck. The inevitable family scramble ensues.

Scientists say our ears and brains are good at locating the source of a sound, but I have my doubts. We use three main processes, experts say. The first is timing. The idea is simple. Sound waves from a single source arrive at the two ears at slightly different times. Suppose a sound is coming from my right. The sound should hit my right ear a few nanoseconds sooner than it reaches my left, setting my hair cells to firing at the same rate as the frequency of the sound. The firing begins slightly sooner in my right ear than in my left, and my brain should be able to detect that difference. This works especially well for low-pitched sounds, but not so well for the higher frequencies. That ringing phone is rather shrill, which may explain why we can't find it.

Another source of directional information is loudness. The idea is simple. If the sound is off to my right, it should be louder in my right ear than my left. Why? Because my head blocks the sound. In the brainstem, one of those "switching stations," called the lateral superior olive, is supposed to detect the difference in loudness between the two ears and give me some directional information. My relative has left the phone under a chair cushion somewhere, so the ring sounds muffled, no matter which way I turn my head. So much for my lateral superior olive.

Another possible method for identifying the direction of a sound may arise from the left–right differences that Sininger and Cone-Wesson have studied. The right and left ears may selectively filter out certain kids of sounds, "preferring" one or the other. The trouble with that ringing phone is that it's high-pitched and rhythmic, much like the music my left ear yearns for. But it's also intermittent and staccato, which my right ear seems to equate with the staccato cadence of speech. Final score on left–right discrimination: nil–nil.

Part of the problem in locating that ringing phone is the echo effect. The sound seems to bounce off the walls, ceilings, and floors in every direction. Another problem is our jabbering while we scramble about, looking for the pesky instrument that will give us no peace. (My relative turns off the voice mail. She doesn't want to miss an important call.) This interference is called masking. It's when the threshold of audibility for one sound is raised by the presence of another (masking) sound. We have learned about this masking effect and taken action to avoid it. We all stay quiet now when the phone rings, because we've given our family member an ultimatum: Whoever loses it finds it.

Chapter 22

Listening and Language

As much as I love science, I love science fiction more. Science tells us what is; science fiction tells us what might be. I recently read a unique and revealing science fiction story, "Immersion," written by University of California at Irvine physics professor Gregory Benford.[1] In his story, Benford invents the technophile's brand of ecotourism, in which visitors to Africa can undertake not a photo safari, but a brain-meld adventure. Benford's characters Kelly and Leon move into the minds of two chimps, experiencing the sensory world from the perspective of a primate species not their own.

Predictably, the humans find themselves trapped in their hosts, encountering hostile rivals and voracious predators, coping with crisis in chimp fashion—their human mental faculties intact, but constrained. While sex, food, and grooming are infinitely satisfying in chimp minds and bodies, Leon and Kelly find themselves frustrated with their limited abilities to communicate. They invent some sounds and signs to get them through, but they long for language:

> It was infuriating. [Leon] had so much to say to [Kelly] and he had to funnel it through a few hundred signs. He chippered in a high-pitched voice, trying vainly to force the chimp lips and palate to do the work of shaping words.
>
> It was no use. He had tried before, idly, but now he wanted to badly and none of the equipment worked. It couldn't. Evolution had shaped brain and vocal chords in parallel. Chimps groomed, people talked.[2]

As Benford's Leon observes through the mind of a chimp, human brains are built for language. They are programmed to speak it and programmed to understand it. Real-life scientists don't question the inevitability of language in the human species, but they want to find out when and how it develops. Is the human brain hardwired for language from birth—or even before? New research on language development is answering a definite maybe.

RESEARCHING LANGUAGE IN NEWBORNS

Brain activity in what will become the brain's left-hemisphere language centers can be detected in infants as young as five days.[3] Behavioral experiments have demonstrated that days- or weeks-old infants can distinguish the "melody" of their native language from the pitches and rhythms of other languages.[4] They can assess the number of syllables in a word and perceive a change in speech sounds (such as *ba* versus *ga*), even when they hear different speakers.[5] Very young babies can also pick up a change of words in a sentence. Two-month-olds can tell the difference between "the rat chased white mice" and "the cat chased white mice" even after a two-minute delay. This ability vanishes when the same sounds are played backward.[6]

"From the first weeks of life," says French researcher Ghislaine Dehaene-Lambertz, "the human brain is particularly adapted for processing speech."[7] Dehaene-Lambertz should know. She's achieved something most other researchers have only wished for. She's persuaded parents to allow functional magnetic resonance imaging (fMRI) studies of their babies' brains, and she's captured clear brain images from the wriggling, mewling infants. The results have convinced her that language processing in the very young relies largely on the same brain circuits that adults use.

Like the rest of the cortex, the temporal lobe (just above the ears) is convoluted in a series of hills and valleys. Three of those "hills," call gyri (gyrus is singular), run parallel to one another horizontally in the temporal lobe. In them most initial speech processing occurs. It begins in the superior (meaning not *better* but *uppermost*) temporal gyrus, which includes both primary and secondary auditory-processing regions. Brain activity in this gyrus increases when we hear speech sounds in any language—whether our own or someone else's—but it's greatest in the left hemisphere when we hear our native tongue.

Numerous studies show activation of the superior and middle temporal gyri when the sounds of speech are processed, as well as activation of the planum temporale, a region of the brain behind the superior temporal gyrus. It lies in the heart of Wernicke's area, which was identified in the nineteenth century as a major center of speech comprehension. Forward from the superior temporal gyrus lies Broca's area, also identified more than one hundred years ago. It is important to the production of intelligible speech. Higher-level language processing, such as interpreting the meaning of complex sentences, gets parts of the frontal lobe involved, as well as the left premotor region of the parietal lobe and the cerebellum.

Language and the Brainstem

Jack Gandour at Purdue University has spent more than thirty years studying language and the brain. What he's uncovered is a river both wide and deep: language engages brain regions in every part of the cortex and employs processing capacity all the way down to the brain's most primitive region, the brainstem. The brainstem is usually described as the area that controls heartbeat, blood pressure, respiration—the fundamental processes of life. It's the brainstem that continues to function in coma patients who endure for years in a persistent vegetative state.

Gandour has found, however, that the brainstem does more than maintain vital signs. It's actually involved in language processing. Gandour and his research team studied brainstem activity in speakers of English and Mandarin Chinese. English is not a tonal language, but Mandarin is (as are most of the world's languages). The meaning of a word depends on the pitch used to speak it. For example, *yi* in Mandarin can mean *clothing, aunt, chair,* or *easy* depending on its melodic contour.

The volunteers in Gandour's experiment heard a series of pitches presented in speech and nonspeech contexts. Speakers of both languages had a stronger brainstem response to language sounds than to the tones, but the Mandarin-speakers exhibited a stronger brainstem response to pitch than the English speakers did, whether the sound was speech or a simple tone. The Mandarin-speakers showed a greater response to sequences of rapidly changing pitches, too.

Gandour concluded that the brainstem facilitates speech processing in the cortex by capturing those parts of auditory signals that are related to language.[8] "Early activity in the brainstem is shaped by a person's language experience, even while the person is asleep," Gandour says.[9] His studies have convinced him that the brainstem plays a greater role in understanding language than most people realize. "Everyone has a brainstem, but it's tuned differently depending on what sounds are behaviorally relevant to a person, for example, the sounds of his or her mother tongue," he says.[10]

Other anatomical differences of the brain suggest that the left hemisphere plays a major role (but by no means a singular one) in the comprehension of language. Various studies[11] have shown that:

- The Sylvian fissure is longer on the left.
- The planum temporale is larger on the left.
- The white matter volume underlying the primary auditory region is greater on the left.

- Pyramidal cells are larger in the left auditory cortex.
- Nerve fibers have thicker myelination in the left auditory cortex.
- The width of the neuronal columns and the distance between those columns are greater in the left superior temporal gyrus than in the right.

Certainly some of those differences develop over time as we learn and use language, but Dehaene-Lambertz's work suggests that the basic circuitry for developing them is already in place in newborn babies. At birth, gray and white matter volumes are larger in the left hemisphere,[12] the Sylvian fissure is longer on the left side,[13] and the left planum temporale is larger than the right.[14] Dehaene-Lambertz studied two-month-old infants, looking at how a nerve bundle—the arcuate fasciculus—was developing. Its maturation, she found, was more advanced on the left than on the right.[15] The arcuate fasciculus runs between Wernicke's area and Broca's area. It's a major pathway for speech, and it develops early in life, long before a child begins to understand or produce speech. Such left-right differences occur in congenitally deaf babies and well as hearing ones,[16] suggesting that the "architecture" is innate and not the result of exposure to sound or language.

Other studies of infant brains during the first year of life show no differences in blood flow in the left and right hemispheres when the infant is resting, in silence. But the left hemisphere becomes more active than the right in response to sounds, especially speech sounds. Dehaene-Lambertz played twenty-second-long speech recordings forward and backward to neonates and three-month-olds while she captured brain images from the babies using fMRI. (The backward recording served as a control. The sounds were the same, but they occurred in a different order, with a different rhythm than natural language.) Both recordings triggered greater brain activation in the left planum temporale than in the right, but the difference was greater still for the normal (forward) speech sounds.[17,18]

Additional activation in parts of the right frontal cortex was seen in awake, but not sleeping, infants who listened to normal speech. The activity occurred in the same brain areas that become active when adults retrieve verbal information from memory.[19]

The French team also found a significant advantage for the native language in the babies' left angular gyrus.[20] In adults, that area increases its activity when subjects hear words, but not when they hear nonsense syllables.[21] Similar increases occur in the area when adults hear sentences in a language they know, but not when they hear a foreign language or backward speech.[22]

Recently Dehaene-Lambertz again used fMRI to measure cerebral activity in three-month-old infants, this time when the babies heard short sentences spoken in their native language. She and her team measured how rapidly responses to speech inputs occurred. They also investigated how particular regions of the babies' brains responded to repetition.

Melodic Intonation Therapy

We can learn a lot about language from those who lose it. Such people have aphasia, or the loss of their ability to speak or understand spoken or written language. Aphasia often results from a stroke that damages or destroys one or more of the brain's language-processing regions. The myriad forms of aphasia range from the inability to generate words for oneself to the inability to understand words spoken by others.

Speech therapists employ numerous techniques to help people with aphasia regain at least some of their language skills. One approach, called melodic intonation therapy (MIT), has shown promise for patients who have trouble making the sounds of speech. The therapy was developed when health professionals noticed that some patients who couldn't speak words could nonetheless sing the lyrics of previously learned songs. MIT can work for people who have lost speech production capacities in the left brain, but who can understand instructions and access their brain's lexicon, the storehouse of words from which meaningful phrases and sentences can be constructed.

The therapeutic course begins with two- or three-syllable phrases, sung in two alternating pitches with contours and stresses that mirror ordinary speech. (Try singing "*Good morning. How are you?*" using a higher pitch for *morn-* and *are* and you'll get the idea.) In rhythm with the singing, the therapist taps the patient's left hand. The theory is that the singing and left-hand tapping engage the right brain, coaxing it to develop language capacities it doesn't have but conceivably could acquire. Later stages of the therapy advance into more complex phrases and musical patterns, with a gradual transition away from singing and back to normal speech.

Therapists who use the technique often report significant progress in speech-impaired patients, if the patients are appropriately selected for the treatment and the MIT program is carefully and consistently conducted. Two patients treated at Beth Israel Deaconess Medical Center in Boston improved their ability to produce meaningful words and to name objects in pictures after forty MIT sessions. MRIs showed significant changes in the brain's right hemisphere.[23] Aphasics don't have to be English speakers to benefit from MIT. Researchers in Iran tried the musical approach with seven patients who met the criteria for MIT. After fifteen sessions, the patients showed improvement in spontaneous speech production, oral expression, and auditory comprehension.[24]

It may be possible to extend the methods of MIT to other kinds of aphasia. Australian researchers worked with four patients whose speech became slurred after traumatic brain injury or stroke. The twenty-four-session, eight-week therapeutic course employed oral motor respiratory exercises, rhythmic and melodic articulation exercises, rhythmic speech cuing, vocal intonation therapy, and therapeutic singing using familiar songs. The patients' speech intelligibility and naturalness improved significantly over time, and the number of pauses they made diminished.[25]

The three-month-olds recognized a repeated sentence even after a fourteen-second interval of silence. The scans showed adult-like activity in the upper region of the brain's left temporal lobe. The fastest responses were recorded near the auditory cortex. Responses slowed down toward the back of the language-processing region and in Broca's area in the left hemisphere. Activity in that region increased when a sentence was repeated, suggesting that infants may be using a memory system based on Broca's area just as adults do.[26]

These findings demonstrate that the precursors of adult cortical language areas are already working in infants even before the time when babbling begins, Dehaene-Lambertz asserts. "The processing abilities of an infant's brain make it efficiently adapted to the most frequent auditory input: speech," she says.[27]

LANGUAGE AND THE HUMAN BRAIN

Benford didn't write about brain research in his science fiction story, but he had it right when he observed that humans yearn to speak and long to listen. Language is characteristic of our species, so much so that we have some built-in, redundant systems. If damage occurs in one of the left brain's speech centers in infancy or early childhood, the right brain takes over, developing normal capacities for understanding and producing language. Deaf people who learn sign language use the same brain regions for signing that hearing individuals use for speech. Broca's area is activated when hearing individuals speak and when deaf people sign. Wernicke's area is used in comprehending both speaking and signing.[28]

At birth, the human brain's asymmetry favors left-brain development of speech capabilities. Although similar asymmetries are seen in other species, no other animal has yet demonstrated a capacity for language. The language-related brain asymmetries observed in human infants aren't found in other primate species that have been studied, including fetal macaques[29] and infant rhesus monkeys.[30] In the great apes, the planum temporale is larger on the left than on the right, although the difference is less pronounced than in humans.[31] Thus, Benford's story, while fiction, sprang from a reasonable premise: language is uniquely and naturally a function of the human brain.

Chapter 23

My Ears Are Ringing

Genius comes in all sorts of packages. Tim Iverson's a genius in an unlikely package—gangly, freckled—an *Andy Griffith Show* Opie stretched into the proportions of manhood, but with his boyishness unaltered. He sits now in my upstairs office, shuffling his feet and mumbling as he figures out why my telephone signal is so weak the answering machine cuts off. Tim's genius is for circuits of the telephonic kind. He probably knows as much about phone lines and the various signals they carry as Stephen Hawking knows about time and space. Tim has previously fixed my telephone, my DSL line, my router, my modem, and a dozen other things mounted on boxes and poles only miles from my house but leagues beyond my understanding.

Tim pulls out a few wires, hooks up some flashing light boxes, and calls a colleague at his home base, asking for signal input. He glances at the pile of paper on my desk, catching a word in a title.

"Tinnitus," he says softly. "I have that."

"You do?"

"Yep, a ringing in my ears. It's there all the time. Annoying," he says, turning his attention back to one of his testing devices. It's buzzing—for real.

After he's diagnosed my problem—in less time than it took me to persuade the dispatcher at the phone company to schedule a service call—we settle in for a chat. I like talking with Tim. He has access to universes of knowledge that I don't, but today he wants to know what I know about tinnitus.

"Is there a pill I can take for it?" he asks.

I explain that a number of drug treatments have been tried and more are being tested, but as of now, there is no pill. "There is, however, a sort of masker you can wear. It creates a sound something like whooshing ocean waves in your ear. You train yourself to hear that instead of the ringing," I say.

"Naw, I don't want to bother with that. I've learned to live with it," he says as he maneuvers his rangy frame out my door.

"Til next time, Tim," I call after him, but he's off, his mind already moving on to the next circuit problem that requires his genius. Tinnitus does not slow him down.

WHAT IS TINNITUS?

People with tinnitus sense a sound in the ears or head that comes from no external source. The sound varies among individuals, often described as ringing, buzzing, chirping (like crickets), whistling, or humming. Most of the sounds are high-pitched and loud. The sounds of tinnitus may be heard in one ear only, both ears equally, or one ear more than the other.

Tinnitus is more common among men than women, and its prevalence increases with age. Nearly 12 percent of men ages sixty-five to seventy-four are affected. African Americans have a lower risk for tinnitus than do whites, and the risk to those who live in the northeastern United States is half that of people residing in the south.[1] It most often affects those who have a hearing loss, but it is sometimes associated with other disorders such as inner ear damage, allergies, high or low blood pressure, tumors, diabetes, thyroid problems, head or neck injury, reaction to certain medications, stress, some forms of chemotherapy, surgery or damage to the auditory nerve, multiple sclerosis, and garden-variety sinus infections.

Twelve million Americans have tinnitus.[2] Most of them, like Tim, simply learn to live with it, but about 1 million Americans have tinnitus severe enough to interfere with their daily activities.[3] Severe tinnitus can induce depression, anxiety, insomnia, and other symptoms that can run the gamut from unpleasant to miserable. With more and more people entering their old age hearing impaired after a lifetime of boom boxes, rock concerts, and airplane noise, tinnitus is becoming increasingly common and increasingly distressing.

A Ringing in the Brain

Although tinnitus if often associated with hearing loss, it is not necessarily a problem in the ears. In many cases, the problem lies in the brain. After an injury, the auditory cortex reorganizes itself in ways that aren't always adaptive.

Neurologist Alan Lockwood at the State University of New York at Buffalo led the first research team to show that tinnitus arises in the brain's auditory-processing centers. In 1999 he reported measures of blood flow in the brains of people who could alter the loudness of their tinnitus at will. He compared their blood flow patterns to those of normal control subjects. He found changes in the left temporal lobe in patients with right ear tinnitus. Stimulation of the right ear produced increased temporal lobe activity on both sides of the brain. He suggested that tinnitus comes not from the ear but from hyperactivity in the brain's sound-processing system, involving a large network of neurons in the brain's primary auditory cortex, higher-order association areas, and parts of the limbic system. Lockwood calls tinnitus "the auditory system analog to phantom limb sensations in amputees."[4]

In 2001 Lockwood studied eight patients with gaze-evoked tinnitus (GET). When GET patients look sideways, the loudness and pitch of their tinnitus increase. The same thing happens to a lesser degree when they look up or down. The condition often develops after surgery to remove tumors near the auditory nerve or the brain's auditory region. Not all GET sufferers have had surgery, however; in some people it develops for unknown reasons.

Lockwood's team mapped the regions of the brain that are active in GET patients and compared the scans with brain activity in normal subjects. The comparison revealed that people with GET have an imbalance between the auditory and visual parts of the brain. In people who don't have tinnitus, looking sideways suppresses the auditory system, almost as if the eyes tell the brain to pay attention to visual inputs, not sounds. But in patients with GET, the visual system fails to suppress the brain's auditory centers.[5] Vision may not be the only cross-wiring that's going on. Some people can change the loudness or pitch of their tinnitus by moving their jaws or pressing on their head or neck. Lockwood thinks the effect of voluntary movements on tinnitus probably comes from abnormalities in the brain's connections between auditory and sensorimotor regions.[6]

Further evidence in support of that idea came in 2008 when University of Michigan researchers showed that touch-sensing neurons in the brain increase their activity after hearing cells are damaged. Susan Shore and her team looked at the cochlear nucleus, the region of the brain where sounds first arrive from the ear, traveling via the auditory nerve. Shore compared normal guinea pigs with guinea pigs that had lost some of their hearing ability due to noise damage. She found that hearing loss altered the sensitivity of neurons in the cochlear nucleus, changing the balance between those that trigger impulses and those that stop them. Shore showed that the some of the neurons actually change their shape, sprouting abnormal axons and dendrites. They may also redistribute themselves, operating in areas where they normally would not.[7]

So taking one source of stimulation away—in this case, sound—triggers neurons that normally deal with input from another source (touch) to change both their structure and their function. "The somatosensory system is coming in (to compensate for hearing loss), but may overcompensate and help cause tinnitus," Shore says.[8]

Treatments

Various treatments have been tried for tinnitus, and some produce better results than others. Some people find that a simple hearing aid works for them. Various forms of counseling and training in relaxation exercises can sometimes help. Acupuncture offers hope to some sufferers, especially those who find that head movement, jaw movement, or facial pressure improve or worsen their tinnitus. People with mild tinnitus who notice the noise mostly at night sometimes get relief from "white noise" machines or "sound maskers." The incoming sounds block the ringing of tinnitus and allow tinnitus sufferers to get to sleep. Wearable maskers are available for daytime use. An audiologist tunes the sound of the masker to match the frequency of the tinnitus sound; the external sound from the masker overrides the phantom sound of tinnitus for some people.

Despite numerous studies on a variety of drugs, no medication has been found effective against tinnitus, and none has been approved by the U.S. Food and Drug Administration for tinnitus treatment. Still, the search for an anti-ear-ringing pill goes on. In 2008 Oregon Health and Science University began clinical trials on acamprosate, a drug traditionally used in the treatment of alcoholism. In preliminary tests it's been shown to reduce tinnitus symptoms, perhaps by affecting the levels of the neurotransmitters glutamate and gamma-aminobutyric acid (GABA) in the brain.

Tinnitus retraining therapy (TRT) is another option. It coaxes the brain to become unaware of the tinnitus sound, much as the normal brain ignores a background noise like a whirring fan or a humming refrigerator. TRT combines counseling with the use of low-level broadband noise generators that patients wear in or behind their ears. TRT takes about eighteen months to complete.[9] The idea is that the sounds of tinnitus don't go away, but the patient learns to live with them. Controlled trials of TRT have achieved mixed results. Scientists in Japan demonstrated improvement in patients after only one month, but more than half of the subjects dropped out of the study, saying they could not tolerate the noise generator they were required to wear.[10]

Researchers at the VA Medical Center in Portland, Oregon, screened over eight hundred veterans to find 123 with severe tinnitus. The scientists randomly assigned the veterans to two treatment groups: half went through TRT and half used tinnitus masking. Both groups showed improvement, but the masking patients tended to show early benefits at three and six months that remained stable over the following months. Those veterans getting TRT improved more gradually and steadily over a year and a half.[11]

Structured counseling is the linchpin of TRT, and the Oregon team also assessed its effects among veterans. They divided nearly three hundred subjects into three groups: educational counseling, traditional support, and no treatment.

What Is TMS?

Transcranial magnetic stimulation (TMS) has been around since the mid-1980s, but it wasn't until the mid-1990s that researchers began to use it and study it in a serious way. TMS uses a specially designed electromagnet that's placed on the scalp. (Not, you can't do this at home with a toy magnet.) The magnet delivers short magnetic pulses to regions of the brain. The pulses are about the same strength as those experienced in an MRI scanner, but pinpointed to a particular area. The pulses travel through the bones of the head, where they can influence the underlying layers of the brain's cerebral cortex. Low-frequency (once per second) pulses may reduce brain activity, but higher frequency pulses (more than five per second) increase brain activation.[12]

Trials are underway to assess the use of TMS in treating depression, chronic pain, speech disorders, drug abuse, posttraumatic stress disorder, Parkinson's disease, and more.

Those who received the counseling attended four group sessions each week. After one, six, and twelve months, the researchers evaluated the severity of the veterans' tinnitus symptoms. The group that received counseling showed greater improvement than the other two groups.[13]

Recently German researchers have begun experimenting with magnetic stimulation, precisely targeted at the brain's most overactive tinnitus regions. In one study, nine patients with chronic tinnitus underwent low-frequency transcranial magnetic stimulation (TMS) for five, fifteen, and thirty minutes. For most, the loudness of their tinnitus was lessened after the treatment—the longer the treatment, the better—and the reduction lasted for up to thirty minutes. Thirty minutes may not seem like much, but the research yielded a valuable piece to add to the tinnitus puzzle. The decrease in perceived loudness didn't come from changes in the brain's primary sound-receiving centers, but from higher up, in the association areas of the cortex where sounds are interpreted and understood.[14]

Longer term use of TMS has produced better, longer-lasting results. In one study, treatment over a five-day period produced significant improvement compared to a sham treatment, in which the patients thought they were getting TMS, but they were not. The improvement lasted for six months according to the patients' self-reports.[15] In 2008 researchers at the University of Arkansas reported success in treating a patient for tinnitus with TMS repeatedly over a period of six months. The researchers found that maintenance sessions helped their patient reduce the severity of his tinnitus, even when it rose again after the original TMS treatment. Positron emission tomography (PET) scans showed that activity in the tinnitus-producing region of the brain fell to normal levels after the long-term treatment, with no deleterious effects on the patient's hearing or cognitive abilities.[16]

So promising is TMS for the treatment of tinnitus that German researchers are now experimenting with various forms of magnetic stimulation applied to different brain areas. Noting that tinnitus patients often demonstrate abnormal activity in the brain's attention- and emotion-processing regions, physicians at the University of Regensburg have combined high-frequency stimulation of the prefrontal lobe with low-frequency stimulation of the temporal lobes. In 2008 they gave thirty-two patients either low-frequency temporal stimulation alone or a combination treatment of both frequencies and both lobes. Immediately after the therapy, both groups reported improvement, but after three months the combined group reported a greater, longer-lasting improvement.[17]

The next time I see my phone-genius friend, I'll suggest TMS treatment for his tinnitus, but I doubt he'll take much heed. He's learned to live with the condition—as have three out of every four who have it.[18]

Chapter 24

Music and the Plastic Brain

Some music teachers are born to perseverance, some achieve perseverance, and some have perseverance thrust upon them. Long-suffering Miss Donahue, my violin teacher, probably qualified on all three counts. Let me tell you about her and her student.

I grew up a poor kid, residing in a cheap, prefabricated house that some unscrupulous developer plopped onto a side street on the rich side of town, while the community's more affluent were looking the other way, failing to recognize the erosion of their property values. Ours was a very class-conscious community, and the ignominy of my lowly position on the socioeconomic pecking order was not lost on me. I felt it every single day, so I tried to keep my head down and my mouth shut—a feat I rarely managed to accomplish.

On the plus side, my lowly status had motivational properties, and I longed to achieve some of the successes that the wealthy girls took so for granted—from their aristocratic bearing (if one can have aristocratic bearing in the second grade) to their mastery of the fine arts. Enter music.

I yearned to play the piano. My fingers skipped across my desktop at school and the kitchen tabletop at home in imitation of what I imagined to be piano playing. I conjured up daydreams of myself, dressed in the latest and most expensive fashions, gliding onto a piano bench to thrill an admiring audience with a masterful performance of something melodious and classical. All the well-to-do girls at my school were performing in piano recitals. I pined to be among them.

Alas, my family had no money either for a piano or for lessons, and there was no room in our small house for a piano. But a teacher in my school contacted my mother and suggested that scholarship money was available for an earnest and capable child like me who might take violin lessons. My mother seized this notion faster than a mantis shrimp traps a hermit crab. She could afford an inexpensive violin (bought on time), and the instrument didn't take up much room. My mother saw me as a violinist—although I did not—so my lessons with violinist Evangeline Donahue began in earnest. I was eight.

Back and forth I trudged to school, carrying my awkward violin case, dragging myself over to the junior high school after hours, when all was dark and quiet, to sequester myself in a tiny room where Miss Donahue attempted to teach me. We labored in a claustrophobic storage room, with shelves of unused textbooks on three walls and high, never-washed windows on the fourth. The windows were painted shut. Dust motes floated effortlessly and endlessly through the dry, hot, still air of what I came to think of as my Tuesday/Thursday prison. But, like a prison cell, my music lesson room had all the necessities. Miss Donahue had a chair and a metronome. I had a music stand, my sheet music, and my violin.

Here's where the perseverance part comes in. Miss Donahue was an accomplished violinist. Sometimes she would gently take my violin from me and play the piece, so I could hear how it should sound. Lovely tones poured from that instrument when she played it. That rosined bow, which moved so effortless for her and so stiffly and uncooperatively for me, released melodies that made my heart sing and my eyes water. Then she would hand the instrument back to me with a soft, "Hear it?" uttered hopefully.

I could not. "Intonation!" Miss Donahue would cry as I fingered and sawed. "Intonation!" she would repeat, perhaps twenty times in every hour we spent together as I tried and tried again to re-create the sounds she had made. Sometimes I thought I had the notes right, but I never did. I longed all the more for the piano where, if one pushed the right key, one reliably got the correct note. Not so with the violin. Even the tiniest difference in the position of a finger produces a pitch too sharp or too flat. I was always one or the other, and I never heard the difference.

Doggedly, Miss Donahue strove for any sign of improvement in her pupil. She heard none, but she persevered. I never progressed beyond stripping horsehairs off my recalcitrant bow and lumbering stiff, pudgy fingers up and down the neck of the instrument. I never even learned to tune it. Close enough sounded fine to me. I did play in some recitals—early in the lineup, so the accomplished players could bring it home for tumultuous applause from the audience before the tea and cookies were served. I had some recital dresses, too, all of them stiff and hot and itchy.

Miss Donahue had her perseverance thrust upon her. She endured me for six interminable years. I still long to play the piano.

The Perfection of Perfect Pitch

Perhaps if I had been blessed with perfect pitch, I might have detected the subtle differences in tone that Miss Donahue expected of me. Perfect or absolute pitch, the ability to recognize a pitch without any point of comparison, was once considered an ability found only in trained musicians. That's because perfect pitch was tested by playing a pitch and asking a person to name it. Since only musicians can name an A-flat or an F-sharp, they were the only ones who could succeed on the test.

Recently, researchers have expanded their testing methods and identified people who lack musical training but can, nevertheless, remember and identify a pitch. Researchers at the University of Rochester's Eastman School of Music and Department of Brain and Cognitive Sciences played groups of three notes in twenty-minute streams to both musicians and nonmusicians. Later the listeners heard the note-groups replayed; their job was to say whether the notes were familiar or unfamiliar. The researchers had a trick in the test. Some of the note groups were played in a different key.

The listeners who had perfect pitch (whether musically trained or not) had trouble with the note groups in a new key. They didn't recognize them. Subjects who relied on relative pitch, however, recognized the note groups as familiar. The test correlated well with the conventional test for perfect pitch in people who were musically trained. More important, it disclosed a small but significant number of nonmusicians who have perfect pitch but never knew it.[1]

Although most researchers attribute the differences in musicians' brains not to genetics but rather to training and practice, the case of perfect pitch may be an exception. It appears that you are either born with perfect pitch or you aren't. The ability is probably inherited as an autosomal dominant trait: if one of your parents has it, you have a 50-50 chance of having it, too.[2]

YOUR BRAIN ON MUSIC

I didn't go to Beth Israel Deaconess Medical Center in Boston to recount the musical miseries of my childhood, but I ended up doing so anyway, because music-brain researcher Andrea Norton was so pleasant, so kind, so approachable —and oh-so-eager for every child to benefit from the gift of music. Mine was a sad saga she thought no child should relive. "Are people like me musically hopeless?" I asked her. "Certainly, music comes more easily to some people, and some people are more naturally musical than others," she answered," but I have never yet met a person that I couldn't teach or a person who couldn't learn." As for teaching music to school-age children, Norton's a big advocate. She thinks music can and should be taught for its own sake, as well as for the benefits it delivers in the academic domains of reading, writing, and arithmetic.

She comes by her opinion honestly. For six years she worked at the New England Conservatory's Research Center for Learning Through Music. She wrote curricula and taught in the public schools to test her approach, teaching young children basic skills through music. First-graders used rhythm to learn to read. Third-graders used melody to learn fractions—all with stellar results. In 2002 Norton joined head neurologist Gottfried Schlaug and his team at the Music and Neuroimaging Laboratory of Beth Israel Deaconess. Since then, she's been working with Schlaug and others on a longitudinal study. Over several years, the researchers have been using fMRI to track children's brain development, attempting to see how the brain changes (or doesn't change) when children study (or don't study) music.

The Beth Israel team started with normal five-year-olds, Norton says, and found no MRI-detectable differences in the children's brains "at baseline," before some of the children went on to take music lessons and others did not. Over the years, Norton and Schlaug have continued interviewing the children and their families, monitoring music lessons and practice time, testing skills, and capturing MRIs of the children's brains at fourteen-month intervals. The researchers' perseverance is paying off. They have begun to see differences. The children who study music are ahead in fine-motor skills such as computer keyboarding and finger-tapping speed, a finding that is not surprising, Norton says, because playing a musical instrument trains the brain's motor system.[3]

But some other findings are less predictable. Norton says the music students tend to score higher on tests of nonverbal IQ, in which they look at a geometric pattern and must choose the correct missing piece. The children who study music "are better at pattern recognition," Norton says, "and they are ahead in vocabulary as well." MRIs show some brain differences, too. The brains of kids who study music are, on average, bigger in one particular region of the corpus callosum, the stalk of nerve fibers that connect the brain's right and left hemispheres. Similar differences have been noted between adult musicians and nonmusicians; a possible explanation is that the larger corpus callosum promotes rapid communication between the hemispheres—a cooperation that's needed, for example, when playing a difficult piano piece.[4] In children, the corpus callosum difference correlates with practice time, Norton reports. The more a child practices, the larger that region of the corpus callosum becomes.[5]

MUSIC AND LEARNING

Norton's team is trying to answer two basic questions. The first is whether training in music changes the brain, and the answer is emerging as an unequivocal

(Music) Lessons from Williams Syndrome

Charlie Betz, age thirteen, has Williams syndrome.[6] Some time when either his mother's egg or his father's sperm was forming, about twenty genes were dropped from chromosome seven. Those twenty might not seem like many out of the 25,000 or so that it takes to make a human being, but they made a big difference in who Charlie is, what he loves, and what he can do. Ask Charlie his favorite thing, and he'll squeal with enthusiasm, "The drums!" Charlie is very good at the drums, and he spends many hours a day practicing. He even plays in a garage band with some of his teenage neighbors.

Not so unusual for an adolescent boy, but Charlie is unusual in some other ways. He can neither read nor write, and he probably never will, his doctors say, although his vocabulary is large and he speaks with clarity and expression. His IQ has been measured at 60, but his social graces show no impairment. He's charming with people—bubbly, sociable, trusting—perhaps indiscriminately trusting—with everyone he meets, and he remembers everyone's face. Charlie can't button his shirt and manipulating a fork at the dinner table is tough, but his motor coordination is excellent when he hits a downbeat. Charlie goes to a special school and loves the people there. When he's not in school or playing the drums, he's listening to music on his iPod. He adores every style from classical to jazz, but not all sounds are pleasurable. He used to cover his ears and cry whenever he heard a car engine starting, but lately he's come to enjoy the sound of a spinning motor. Some sounds that don't seem especially loud to other people—such as the whoosh of the vacuum cleaner—are too loud for Charlie; he covers his ears and runs away. Other sounds—such as the whir of the kitchen blender—can keep him fascinated for hours.

Neuroscientists have studied the brains of children like Charlie is hopes of better understanding not only Williams syndrome, but also the structure and development of the normal brain. Compared to control-group children, children like Charlie tend to have a small volume when both overall brain size and the size of the cerebrum are measured, but the volume of the cerebellum is no different, although the brainstem is disproportionately small. Relative to typically developing children, Charlie probably has less white matter in his cerebrum and less gray matter in his right occipital lobe.[7] (The gray matter is composed primarily of the bodies of neurons. It's found mostly in the brain's outer layer, the cerebral cortex. The white matter lies beneath the cortex. It's made mostly of axons, thin projections from neurons that carry impulses away from the cell body.) In Charlie's primary visual cortex, the neurons are smaller and more densely packed than they should be—perhaps reducing the number of connections that can be made between and among them. In the primary auditory cortex, it's the opposite: cells are larger and more loosely packed, perhaps allowing for more connections.[8]

Charlie's brain is convoluted into folded gyri and sulci, but not everywhere in the normal way. The central sulcus, the large deep groove that runs across the top of the head separating the frontal and parietal lobes, is abnormally shaped, as if it failed at some point during development to twist back on itself.[9] In the right hemisphere, the Sylvian fissure (also called the lateral sulcus), which is a prominent valley in the temporal lobe, cuts horizontally, and fails to ascend into the parietal lobe the way it should.[10] Studies have also revealed an abnormality of the planum temporale, a region of the auditory cortex known to play a role in musical learning and processing. In most people, it's larger in the left hemisphere than in the right, but in children like Charlie, it's the same on both sides, because the right is greatly expanded.[11]

Differences are seen, too, when fMRIs reveal patterns of activity while listening to music or noise. In children with Williams syndrome, the brain regions that "light up" in normal children (such as the superior temporal and middle temporal gyri) don't always respond to either music or noise. But children like Charlie tend to produce less brain activation in the temporal lobes along with greater activation of the right amygdala—a part of the limbic system intimately involved with emotion. What's more, children like Charlie show a much more widely distributed network of activation while listening to music, from the higher cortical regions all the way down to the brainstem.[12]

Perhaps this greater pattern of brain activation explains Charlie's love of music. Perhaps auditory processing and sociability are somehow related to the same set of twenty genes, although what the causal pathway might be is anybody's guess. Charlie's not bothered. He loves music for its own sake and for the joy that it brings to his life and the lives of others. Maybe in that respect, Charlie's not so different after all.

yes. Study after study has confirmed differences between the brains of trained musicians and the rest of us (who never learned to tune our violins). In one such research project, fMRI was used to compare brain activity in professional violinists with that of amateur players while they performed the left-hand finger movements of a Mozart concerto. The brain scans showed many similarities, but some important differences, too, especially when it came to efficiency or economy of effort. Brain activity in the professional players was more intense and more tightly focused spatially. It centered in the primary motor cortex on the right side of the brain, the region that controls the left hand. In the amateurs, activity patterns were more diffuse and spread into both hemispheres. The supplementary and premotor regions of the cortex were more active in the amateurs. Those same areas are known to be active when complex skills are being learned but have not yet become automatic.[13]

In a study published in 2003, Norton's team leader Gottfried Schlaug and a German colleague, Christian Gaser, compared gray matter volumes in the brains of three groups: professional musicians, amateur musicians, and nonmusicians. The researchers found a correlation. The better trained and more skilled the musician, the greater the volume of gray matter in a number of regions, including the primary motor and somatosensory areas, the premotor areas, and parts of the parietal and temporal lobes, as well as in the left cerebellum (automatic skills) and the left primary auditory cortex (sound processing).[14]

The second question is harder to answer. It's whether musical training can facilitate learning in other domains. Because of her experience teaching reading and math using music, Norton believes that it can. But objective evidence for such crossover abilities—from music to language or computation—is hard to come by. Nevertheless, support is mounting. Several studies have shown that training in music enhances visual-spatial performance, a set of skills important to—among other things—mathematics.[15] Music training also appears to enhance coordination between visual perception and the planning of related actions.[16] Music training enhances verbal memory in both adults[17] and children.[18] Some studies suggest that music training can even effect significant, albeit modest, increases in IQ.[19] Consistent with Norton's classroom experience, rhythm-based music training has been shown to help children with dyslexia become better spellers.[20]

Norton offers some plausible explanations for how musical training might enhance other domains of learning. Spatial skills may be developed through musical training because musical notation is itself spatial, with relative pitches denoted by a series of lines and spaces. Mathematical skill may grow because the rhythm of musical notes is inherently a mathematic relationship of proportion, ratios, and fractions. For example, a half note is sustained twice as long as a quarter note. Spelling and reading might improve because both music and language require the brain to break streams of sound down into smaller perceptual units and to find meaning in the pitch—a skill that's required in inferring meaning from the melody of language. An example is the questioning attitude revealed when the pitch rises at the end of what might otherwise be a declarative sentence.[21] (To hear the distinction, try reading these sentences out loud: You've come to see Mary. You've come to see Mary?)

Norton convinces me that musical training does some very fine things for the brain, although how much training is required and how apt the student needs to be remain a mystery. But evidence is accumulating that positive effects are significant and they don't stop in childhood. For example, researchers a Vanderbilt University asked matched groups of college students—half music students and half nonmusic students—to dream up novel uses for common household items. The musicians suggested a larger number of uses, and their ideas were more creative. Brain imaging using near-infrared spectroscopy (NIRS) measured oxygen flow to brain regions. While the nonmusicians used primarily their left hemispheres during the task, the musicians used both the left and the right. While it's possible that the students who studied music were better divergent thinkers to begin with—and maybe better cross-brain thinkers, too—the researchers offer a different explanation. They suggest that learning to use both hands when playing an instrument may explain why musicians use both hemispheres more and do better on a task that requires divergent thinking.[22]

I end my visit to Beth Israel Deaconess a little wiser about my youthful failures. I never became a violinist, but I became a voracious reader and an indefatigable writer. Maybe I learned something from Miss Donahue after all, if nothing more—and nothing less!—than perseverance.

Chapter 25

Cochlear Implant

One Man's Journey

Stephen Michael Verigood is a twenty-seven-year-old deaf poet who lives in Aliquippa, Pennsylvania, about twenty miles north of Pittsburgh. He lives alone, and his family is spread "all over," Stephen says, in Georgia, Ohio, and Arkansas. He's the author of three poetry collections and "a very scratchy golfer." He designs his own Web site, featuring his photographs and poetry.

Stephen was born completely deaf, but he was good at reading lips so his condition went unnoticed until he started preschool. He was fitted for his first hearing aid when he was four years old. "With the hearing aid," Stephen says, "I was only able to hear the things that were necessary to get through life, going through mainstream education with a speech therapy class every week." His hearing aid worked for him for more than twenty years. Then . . . I'll let Stephen tell the story in his own words:[1]

"I was working as a salesman at Macy's, helping a customer, when my hearing aid started making a weird vibration in my ear. This had happened before, but this time the buzz wouldn't go away. Over the next two weeks, the din got to the point where I couldn't handle it anymore. I couldn't function in the work force without my hearing aid, so I knew I had to look for a solution. I was afraid that I had a brain tumor and it was affecting my ears. I was afraid that I was going to have to be on disability for the rest of my life. I was only twenty-four. I still had a lot of living to do.

"I went in for testing. There was no brain tumor, but no one could give me a reason for the buzzing of my hearing aid either. The doctor suggested a cochlear implant. I was against the idea. I had done some research on implants when I was in high school, and I didn't like what I'd learned. I thought I was too old for one and that if I did decide to get one, I would have to start over again learning sounds, even speech therapy. I did not want to do that, but I knew I needed to keep an open mind, so I did some new research—this time with a clearer insight and a fresh eye for details. I saw that I had been too quick to judge and that the technology had improved since I was in high school. I decided to try an implant after all.

"What followed was nearly a year of tests, paperwork, referrals, more tests, more paperwork, and waiting—the interminable waiting. The problem was that I had no health insurance, and even when I finally became insured through a government program, the insurance people rejected me for the procedure, saying cochlear implants were not covered. But the cochlear implant specialist at Pittsburgh Ear Associates (located at the Allegheny General Hospital in downtown Pittsburgh) appealed on my behalf—attesting that an implant was a medical necessity for me. Wonder of wonders, she won the appeal! The date and time for my surgery were set: July 16, 2007, at five o'clock in the morning.

"At first, I was elated, but about a week before my surgery date, I started to have serious doubts. I was organizing a poetry reading to be held at my church the night before my operation, and I was thinking hard about canceling the reading and the surgery, too. I was wondering if I was ready for what was to come. I kept trying to imagine all the wonderful things that would happen once I could hear, but my excitement invariably turned to nervousness. *Was I ready?* I didn't know. I realized that life wouldn't be the same, but if I didn't go through with the surgery, I would be letting a lot of people down. In the end, a simple message of support from a woman I love pushed the clouds of doubt away.

"The poetry reading went off without a hitch. After reading a few poems and telling the stories behind them, some friends of mine sang a song. While they were singing, I kept imagining what the song would sound like once my implant was turned on. Surrounded by family and friends, my nerves faded fast, but I tried to sleep that night and I couldn't. As my father and I were driving to the hospital, my stomach was in knots. This was going to be only the second surgery in my lifetime, and the first one took place when I was too young to remember. I think it was a hernia, but I cannot recall.

"The morning of the surgery, I thought I'd never get through all the forms, questions, waiting rooms, and lines. There were dozens of people there, waiting just as I was, and I wondered about their stories. *Were they getting implants, too?* After changing into a hospital gown (open in the back!) and climbing into a bed, I had to answer still more questions from a nurse who was carrying a huge folder. I felt like a suspect in custody. After a good ten minutes of questions, she told me it was time. I took off my hearing aid and handed it to my father. The nurses wheeled me out of the room and down a long hallway. I felt like I was being paraded down Main Street in Aliquippa. They pushed me into a room that felt like a morgue. Ten or fifteen other patients lay on gurneys there, and they all had names on the fronts of their gurneys. I couldn't hear a thing, but somehow I knew what people were saying.

"My surgeon stopped by my bed and shook my hand. He told me that everything was all set and that they were just waiting for the drugs to kick in. *What drugs?* I looked at my arm and saw a needle sticking out. *Oh,* that *drug. The one to knock me out.* I was so nervous about the surgery, I hadn't noticed that they had placed a needle in my arm. The last thing that I remember was being wheeled into a smaller room. I saw the surgeons and a few nurses. They packed me in tight wrap and placed my head straight up. The nurse looked at me and, from reading her lips, I know she said, 'Think of something good.'

"I woke up in what felt like five minutes. I saw my father standing over me, and then I saw on the clock that it was nearly two in the afternoon. My head was throbbing, and a huge bandage covered my ear. It looked like an athletic cup with a strap around my head. The worse part was that it felt as if there were a little man standing inside my ear, blowing a whistle as loud as he could. It was nearly five o'clock before I could be discharged. They brought me a wheelchair, and I wanted my father to push me as fast as he could through the halls. I guess the drugs hadn't worn off yet.

"I cannot recall anything about the drive home. I do remember that once I got to my grandmother's house, I went straight to my father's bedroom and fell on the bed. I was out for the rest of the day. When I woke up, I think it was close to eight or nine that night, and my father had already been to the pharmacy to get my prescriptions. He gave me directions on how and when to take the pills. I didn't pay any attention to his directions. I just took the pills. The little guy was still blowing his whistle in my ear. The pills made me tired, so I went back to bed, and I didn't wake up until the next morning.

"The next day, the surgeon called my dad to follow up on me. I asked if the ringing in my ear was normal. He told me that it was, and that it would go away in a couple of days. (It did.) He also said to keep the 'cup' on my head for one more day. In between the times that I was asleep and the times I was being miserable around the house, things were not that bad, but I couldn't open my mouth. I tried slurping milkshakes and soup, but I couldn't get much down. Because I couldn't eat, I lost a lot of weight, most of which I gained back later.

"Two nights later, I went home to my apartment and emailed everyone that I knew to tell them that the surgery was a success. Not too long after that, I received about two hundred emails, all wanting to know how and what I was feeling. Some people wanted to see pictures of my face and ear. I took out my camera and did my best to capture those images. I posted them on my Web site. *Not pretty!*

"The next afternoon, I got restless and tired of holing up in my apartment, so I decided to go for a drive. I went to Macy's to see coworkers I hadn't seen in a week. To cover my weird surgical haircut, I wore my Steelers 'Gilligan' hat. My eye was swollen, as if I had survived a fight with Rocky. I must have looked pretty weird walking around Macy's talking to people.

"That little jaunt took more out of me than I realized. When I got home, I sat on my couch and turned on my television. I took my pills and before I knew it, I was asleep. When I woke, it was getting close to three in the morning. I had slept for ten hours. I went into the bathroom and saw that the swelling on my face had gone down. I took my pills again and went back to sleep. I was out until the next afternoon. I spent the weekend with my grandparents. I told them that I was getting bored and I wanted to get back to work. My grandfather, the wisest man I know, told me to rest. 'It's not every day your head gets cut open and operated on,' he said.

"As I began to feel better, the next step of the journey was to be 'turned on.' That was when the external device of the cochlear implant was connected to my head and activated. It would be another month before that could happen, so there I was . . . waiting again. On Tuesday, August 14, 2007, my wait ended. All through the day, I felt jumpy and anxious. Some friends went to the hospital with me, and we borrowed a video camera so we could record everything.

"It was not a foregone conclusion that my implant would work. I had heard stories about people who went through the procedure, only to find that the device did not improve their hearing or that they did not like the sounds they heard once it was turned on. I sat nervously as the implant specialist put me through the paces. She played a series of beeping sounds and I had to acknowledge when I heard them. When I had taken that test before, I had strained just to pick up that beep. Now, noticing the sound was almost effortless.

"My implant came with four software programs, I learned. It was up to me to choose which program I wanted to keep. The implant specialist gave me the basic program that allowed me to hear almost everything. She also gave me another program that lets background noise fade away in a noisy situation. She made copies of the two programs and told me to try them at home and figure out which kind of environment I was going to be in most often. She told me to stay off the telephone because I was not 'ready' for that step.

"After the session was over and we left the office, I kept looking around trying to place where each sound was coming from. It was a very noisy world, as I was only beginning to find out. We hit rush hour traffic leaving Pittsburgh. I rolled down the window and heard a car radio playing in the vehicle behind us. When I got back to my apartment, I watched the video we had just shot. Amazingly, I was able to hear and understand everything that was said. After I finished the video, I sat on my couch and just listened. I began to hear things I had never heard before: creaking noises, running water, people talking next door, the phone ringing. I did not answer it. I took it off the hook. I did not want to tempt myself into doing something the doctor had told me not to.

The Brain and the Cochlear Implant

Stephen's implant is the Nucleus Freedom model, manufactured by the Australian company Cochlear, Ltd. His implant was surgically inserted just under the skin behind his right ear. An electrode array extends from the implant into the cochlea (see Chapter 21). Behind his ear, Stephen wears an external sound processor with an attached coil. The sound processor converts sound waves to digital code. The code is then transmitted via the coil to the implant. The implant changes the digital code to electrical signals and sends them along the electrode array. The electrical signals stimulate nerve fibers in the cochlea to trigger an impulse, transmitting sound signals to Stephen's brain. Stephen is one of 120,000 people worldwide who've received one of the company's implants.

Not all do as well as Stephen, but the device is seldom to blame. Ironically, the brain's remarkable plasticity—which is so beneficial when recovering from a stroke or a paralyzing injury—is at fault. Children born profoundly deaf do best if they have implants before age four. Any later and the auditory cortex may never fully develop the neuronal connections needed to process sound inputs. It's a sort of "use it or lose it" situation. The brain's sound-processing centers can mature properly only if sound impulses trigger them to do so. Stephen thinks he adapted easily to new sounds after his implant because he was ready for sounds, having experienced some through the hearing aid he had worn since childhood.

To find out how brains like Stephen's adapt, British and German researchers used positron emission tomography (PET) to study brains at work when they process syllables, words, and environmental sounds. The researchers compared people who had received cochlear implants after a period of profound deafness with normal hearing subjects. Both groups used the left and right superior temporal cortices and the left insula (at the back of the temporal lobe) to process sound. Wernicke's area responded to speech sounds in people with normal hearing but was not specialized in those with cochlear implants. The subjects with implants recruited additional regions to process speech, especially parts of the left occipital lobe, which is usually a visual-processing area of the brain.[2]

"Later, I went to my grandparents' house. It was getting dark, and dew was forming on the grass. I took a walk through the yard and, for the first time, I heard my feet slosh through the wet grass. My uncle, ever the prankster, lit a firecracker and threw it. The loudness of the explosion made my head so dizzy, I had to grab a chair on the deck to keep from fainting.

"As I got ready to go back to work at Macy's, I began to get nervous again. After working with these people for six years, I was going to hear their voices 'naturally' for the first time. I dawdled. I wanted to be the last one to arrive at work. I wanted everyone to be working already when I walked in, so I could avoid being the center of attention. At the store I headed straight for my team's room, but a coworker stopped me to ask, 'Did it work?' I heard her and understood her. I nodded and she gave me a big hug. Then it started. People gathered around, asking questions. I made it a point to walk around the store, finding people to hear and to answer their questions. They had a lot of questions. The most annoying one was, 'Can you hear me now?'

"*Yes, I can. I can hear you now.*

"By the end of my day, I wanted to escape. I wanted to get the device off my head. Everything was just too loud. My head was spinning, and I wondered how I was going to make it home. I left the store and walked to my truck in the parking lot. I heard my feet pound the pavement and the starting of many cars. My eyes hurt and I was getting a migraine. I made it home and turned the device off. The silence was so welcoming. *Now I know what they mean by peace and quiet.*

"The next morning, I turned the device on again. Morning sounds flooded over me, and I feared an anxiety attack. I plopped onto my couch and sat there, staring into the blackness of the television screen. I asked myself, *Why did I go through all that? What am I going to hear from now on? How are people going to treat me? Will this change what my poetry is about? Am I going to be a different person? Will I be accepted now that I can hear? What is going to happen to me? Will the sounds get any better? Did I do the right thing?*

"When I snapped out of it, I looked at my clock and saw that I was going to be very late for work. People would be worried about me. I took a deep breath and left my apartment. When I got to Macy's, I went around to tell my team members that I was okay—that I had only needed to think about some things. The rest of day went as normally as normal can . . . and it's been that way ever since."

PART SIX

BEYOND THE BIG FIVE

Reality is merely an illusion, albeit a very persistent one.
—Albert Einstein

Chapter 26

Synesthesia

When Senses Overlap

One of my favorite pieces of music is Ferde Grofé's *Grand Canyon Suite*. Hearing that music seems to transport the listener to the canyon. The melodies and rhythms evoke mental images of majestic cliffs, rocky trails, and plodding donkeys hauling heavy packs from the canyon's rim to the muddy Colorado River far below. When Grofé spins up an orchestral lightning storm, the listener can almost feel the rain.

Almost.

For most of us, Grofé's music is simply bringing back memories. We have experienced—either in real-life or on TV—the clip-clop of donkeys and the roar of thunder. It is easy for us to relate these sounds to visual memories.

But for a small number of people, listening to music is an entirely different experience. Some people see sound as color. Others get taste sensations from musical sounds. For one professional musician, a major third tastes sweet, while a minor third is salty.[1]

Hearing isn't the only sense that can overlap and meld with other senses. Some people perceive numbers as shapes, words as tastes, or odors as tactile sensations in the hands. People who experience such an overlapping or intermingling of the senses are called *synesthetes*, a term that means literally "those who combine the senses." The phenomenon itself is called synesthesia.

A CASE STUDY IN SYNESTHESIA

"Jerry Springer changed my life," says Karyn Siegel-Maier.[2]

"I'm fairly normal (ahem), forty-seven years of age, married with three grown sons and I live in the Catskill Mountains of New York," she relates. "I've scratched out a living over the years as a singer, bookkeeper, legal assistant, Web master, online community manager, certified software consultant, balanced life coach, and freelance writer and editor." Karyn has written hundreds of magazine articles and authored two books published in four languages. Her specialty is herbal medicine.

All her life she has experienced letters as colors. The association is so strong that Karyn can easily visualize any letter or word in its associated colors. "For instance," she explains, "the letter *d* to me has a dark, brownish-red color, sort of like burnt sienna. The letter *o* is an off-white color. The letter *g* is a dull yellow-golden color." When Karyn perceives a word made from those letters, *dog*, it's a lighter shade of sienna because the first letter of the word dominates the color of the word in its entirety. Her name, she says, is a deep maroon. *Thunder* is brown. *Ball* is a pale grayish-blue.

Karyn explains that she does not actually see letters and words in color on paper. Black type is black type. Only when she hears or thinks of a word does the color image flood her mind. "The interesting thing is that if you asked me what color I associate with any particular letter, word, or number today, I'll consistently give you the same answer, whether it's two weeks from now or two years." Karyn says she sometimes has difficulty explaining the color of a letter because there is no word for a color like greenish-gray-white. But she can easily pick the right color from a palette or color wheel.

Back to Jerry Springer.

Karyn says she watched an episode sometime in the 1980s, back when Springer's program was "a talk show with a panel of guests who weren't inclined to tear off their clothes and beat the crap out of each other. In fact, Jerry's show introduced very serious topics," Karyn explains. "One day, I was flipping channels when I saw that he had guests who were talking about synesthesia. I had never heard the term before, but was stunned when these people described tasting shapes. . . . One guest said he could see words in color. I called my ex-boyfriend-now-friend on the phone and told him to turn on the show. So there. Now he gets it!"

Karyn says her synesthesia is not a hindrance in any way, "[but neither] will it help me win a Pulitzer Prize!" Color does, however, influence her daily life in a big way. She's always had an artistic bent, and she describes herself as once a "fairly good artist with pastels. . . . At one time, I was in possession of high-quality French pastels that numbered more than three hundred shades. For someone with synesthesia, that is drool producing."

Her personal environment is important to her, and she's an excellent decorator. She paints walls in colors that make most people nervous until they see the final result. "My entire home is decorated in early American primitive, with the primary colors being black, barn red, hunter green, mustard, and colonial blue. I'm not afraid to mix those colors in the same space. In fact, my home is an experience in visual stimulation," she says.

WHAT IS SYNESTHESIA?

Karyn isn't the first synesthete, and she won't be the last. Synesthetes have been around throughout human history, and their talents have occasionally been noticed and described. The nineteenth-century French poet Baudelaire captured synesthetic experiences in his poem "*Correspondances*," meaning associations. To Baudelaire, echoes mixed together as shadows. Some smells were the sound of an oboe; others were the green of meadow grass. A similar synesthetic portrayal came from another French poet of that same era, Arthur Rimbaud. In his "*Le Sonnet des Voyelles*," he associated each of the vowels with a color: *A* was black, *E* was white, *I* was red, *O* was blue, and *U* was green. Experts aren't sure whether Baudelaire and Rimbaud were genuine synesthetes or if they were simply writing in metaphor.

Most of us use synesthetic metaphors in everyday communication, scarcely thinking about what we're saying. When my friend Ida proclaims she's "all points and plugs today," her meaning is clear: she's feeling out-of-sorts, edgy, nervous. While Ida relates emotions to shapes and objects, other common comparisons use colors for mood states: "I'm feeling blue," for sadness; "I saw red," for anger; and "She was in a brown study," for contemplation. We often use synesthetic phrases to describe objects or events. Examples include "loud suit," "bitter cold," or "prickly disposition." Old whiskey tastes "smooth" and wild women are "rough."

But true synesthesia is much more than figurative language. The combining of senses is real for those who experience it. Some synesthetes see sounds as colors. Others perceive sounds or tastes as shapes felt by the skin. In his book *The Man Who Tasted Shapes,* Richard Cytowic tells of his friend, Michael, who wouldn't serve a chicken dish until it had enough points on it. To Michael every flavor had shape, temperature, and texture. He felt and sometimes saw geometric shapes when he tasted or smelled food. Flavors produced a touch sensation:

> "When I taste something with an intense flavor," Michael [said], "the feeling sweeps down my arms into my fingertips. I feel it—its weight, its texture, whether it's warm or cold, everything. I feel it like I'm actually grasping something." He held his palms up. "Of course, there's nothing really there," he said, staring at his hands. "But it's not an illusion because I feel it."[3]

Synesthetes typically experience their "joined" senses for many years before discovering that others do not perceive the world as they do. Karyn wasn't aware of her synesthesia until she was in her early twenties. She had learned to dismiss it years earlier. "That's probably because I realized that other kids didn't have the same perception when I was in grade school, so I just didn't talk about it for fear of getting labeled," she explains.

She confided her secret to her college boyfriend. He felt certain that her association of colors with letters stemmed from cards she might have used to learn the alphabet as a child. Karyn disagrees. She recalls the alphabet posted above the blackboard in the classroom: dark green cards with white letters. "The funny thing is that this boyfriend has remained a close friend for the last twenty-some years, and he's tested me a few times. What color is the letter W? It's greenish-gray-white. Same answer, every time. Give me a pack of crayons and I'll color *War and Peace* the same way today as I would ten years from now. This leads me to believe that this color association is innate for me, not learned or random. I just do it without thinking and without the slightest hesitation."

How many of us experience some of our senses in combination? Cytowic estimates that one in every 25,000 individuals "is born to a world where one sensation involuntarily conjures up others," but he suspects that figure is too low. Hearing words, letters, or numbers as colors is the most common form of synesthesia. British scientists suspect that as many as one in every 2,000 people can do it.[4] Professor Joel Norman of the University of Haifa thinks that perhaps one person in ten has some synesthetic ability without knowing it.[5]

Although researchers have long believed that synesthetic experiences are unique to each synesthete, and not generalizable across all synesthetes, recent evidence suggests that synesthetes may actually share some commonalities—at least as far as color-letter synesthesia is concerned. Psychologists Julia Simner of the University of Edinburgh and Jamie Ward of the University of Sussex found that synesthetes often share the same associations. For example, the letter *a* frequently evokes the color red, while the letter *v* is often purple. The common letters tend to pair up with common colors, while less frequently used letters induce rarer hues. The researchers think their studies show that synesthetic match-ups are at least partly learned.[6]

No one knows how synesthesia works, but the best guess is it comes from some kind of cross-communication between adjacent areas in the brain's cortex.[7] Evidence also suggests that synesthetes

- Inherit their cross-sensory abilities from their parents.
- Are more often women than men.
- Are often left-handed or ambidextrous.
- Have normal nervous systems and mental health.
- Are often highly intelligent and have excellent memories.
- Sometimes experience certain learning or thinking difficulties such as a poor sense of direction or trouble with math.

ARE WE ALL SYNESTHETES?

Studies suggest we may all be synesthetes of a sort, whether we realize what's happening or not. Why? Because the senses influence each other in the "first-stop," stimulus-processing regions of the brain, as well as "higher up," where sensory information is integrated. This was demonstrated in one study in which researchers at the University of California at San Diego presented volunteers with a flash of light, pulsed in between two brief bursts of sound. Although the light flashed only once, many subjects reported seeing two flashes of light, one after each sound. Event-related potentials, which are recordings of electrical activity in various brain regions, showed that those people who saw the illusion experienced a spike in a particular waveform in their visual cortex within thirty to sixty milliseconds after hearing the second sound. That spike by itself, however, couldn't predict seeing the phantom flash, but a spike of a different wave type occurring in the auditory cortex even earlier, at twenty to forty milliseconds, could. Thus, the investigators attributed the illusion to extremely rapid interaction between the brain's hearing and visual circuitry.[8]

In another study, Canadian researchers used magnetoencephalography (MEG) to study changes in the brain that occurred after volunteers viewed famous and not-famous faces or heard famous and not-famous names. It was no surprise that famous faces produced big changes in the brain's visual areas—more so than not-famous faces did. The same is true of activity in the auditory-processing region for famous versus not-famous names. But it was the crosstalk of the two senses that proved most interesting. Famous faces triggered activity in the *auditory* region; famous names caused the *visual* areas to rev up. The brain doesn't need much time to make these associations and serve up an enhanced sensory perception of a known quantity—something on the order of 150 to 250 milliseconds.[9]

Such pseudosynesthetic experiences occur, in part, because different sensory signals provoked by the same object or event are integrated in other areas of the brain besides the primary sensory areas. Such multisensory, integrative regions include several areas of the parietal, temporal, and frontal cortex, as well as underlying regions, including the superior colliculus, the basal ganglia, and the putamen.[10] Information exchange occurs between and among those multisensory areas and the primary-processing regions. The integrative areas send feedback to the individual sensory-processing regions and influence their continuing action. This can result in illusions such as the ventriloquist effect, in which the location of a sound can be perceived as shifting toward its apparent visual source; and the McGurk effect, in which viewing lip movements can change how speech sounds are heard.[11] Other examples include the following:

- A research team at Oxford University scanned volunteers with fMRI while the subjects smelled odors, viewed colors, or did both at the same time.

Closet Synesthetes

There's no doubt in my mind that some people are synesthetes, but they never know it. One who comes to mind is Septimus Piesse, the nineteenth-century French chemist who thought that smells affect the nose just as sound affects the ear. For Piesse, smell was a musical thing, and he plotted odors on the same kind of scale musicians use. The odors that Piesse considered "heavy" he assigned to the bass clef. Sharp, pungent odors he placed high on the treble clef. He wrote in 1857:

> Scents, like sounds, appear to influence the olfactory nerve in certain definite degrees. There is, as it were, an octave of odors like an octave in music; certain odors coincide, like the keys of an instrument. Such as almond, heliotrope, vanilla, and orange-blossoms blend together, each producing different degrees of a nearly similar impression. Again, we have citron, lemon, orange-peel, and verbena, forming a higher octave of smells, which blend in a similar manner. The metaphor is completed by what we are pleased to call semi-odors, such as rose and rose geranium for the half note; petty grain, neroli, a black key, followed by fleur d'orange. Then we have patchouli, sandal-wood, and vitivert, and many others running into each other.[12]

I'd also nominate for synesthete-hood the modern-day scent maverick, Luca Turin (see Chapter 7), whose olfactory sense has to be something

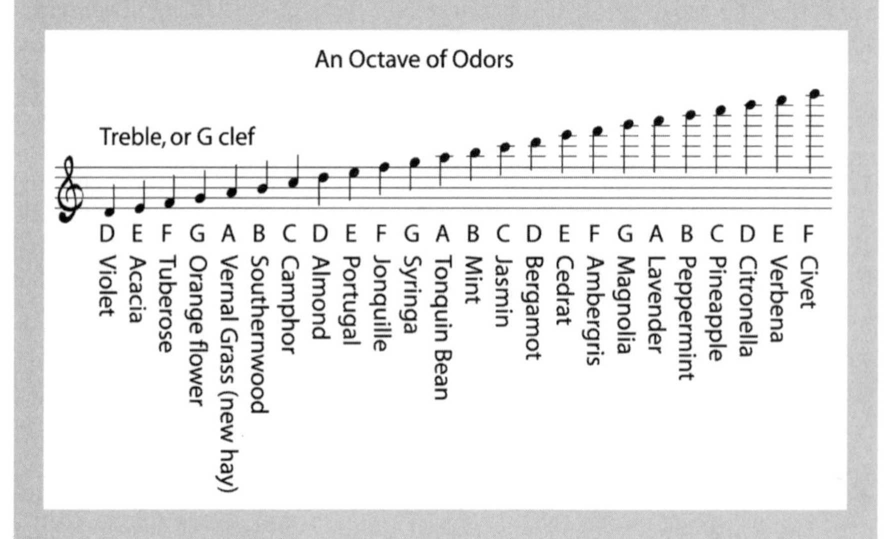

superhuman—an assertion he vehemently denies. Protestations aside, Turin describes scents in ways that defy mere mortal comprehension, and his descriptions often possess synesthetic attributes. For example, Turin told Chandler Burr, while discussing scents: "I've always thought that esters, fruity, are Mozart. The melon notes—helional, for example—strike me as the watery Debussy harmonies, the fourths. Beethoven in his angrier moments is quinolines, which you get in green peppers."[13]

Spoken like a true synesthete.

They found that activity in caudal regions of the orbitofrontal cortex and in the insular cortex increased in proportion to what the people perceived as better and better "matches" between the olfactory and visual cues.[14]

- A different Oxford team asked volunteers to judge the roughness of abrasive surfaces while they heard a variety of sounds. High-pitched sounds made the surfaces feel smoother.[15] When subjects smelled lemon and felt fabric swatches, they judged the fabrics as softer that when an animal-like odor accompanied the touching.[16]
- Wine tasters in Bordeaux described the "nose" of a white wine and the same white wine that had been colored—without their knowledge—with an odorless red dye. They perceived the white wine as have a red wine smell (described in terms such as prune, clove, cherry, cedar, musk, violet, and cinnamon) when it was colored red.[17]

All of us—true synesthetes or not—have to admit that we have little control over our perceptions. Synesthetes can no more stop themselves from hearing colors or tasting shapes than the rest of us can block out the music of an orchestra or the flavor of chocolate. Seldom do synesthetes describe their commingling of the senses as a handicap. In fact, quite the opposite. "I think the color association helps me on several levels, including in my profession," Karyn says. "First, being able to 'see' words in color allows me to recall information that I've read without necessarily having to memorize facts. That was very helpful in college when I was taking a lot of earth science courses. That kind of recall is also useful when encountering unfamiliar words or terms. By simply 'seeing' the first few letters of the word or term in color, I can then assign a 'feeling' to the word, by the 'warmth' or 'coolness' the colors represent. In short, the crazy thing has helped me to develop somewhat of a photographic memory. . . . So it might be fair to say that my synesthesia has helped to expand my creativity and self-expression. Or, it could be the other way around: the creative processes in my brain have led to seeing letters and words in color. Who knows?" Karyn asks. "Maybe it's both."

Chapter 27

If This Looks Like Déjà Vu, You Ain't Seen Nuthin' Yet

Sudden Light
I have been here before,
But when or how I cannot tell:
I know the grass beyond the door,
The sweet keen smell,
The sighing sound, the lights around the shore,
You have been mine before,-
How long ago I may not know:
But just when at that swallow's soar
Your neck turned so,
Some veil did fall,- I knew it all of yore,
Has this been thus before?
And shall not thus time's eddying flight
Still with our lives our love restore
In death's despite,
And day and night yield one delight once more?
—Dante Gabriel Rossetti (1854)[1]

Déjà vu is like love. If you've experienced it, you know it. If you haven't, it's a mystery.

Most of us fall in love at least once in our lives. So, too, with déjà vu. Studies say 70 percent of us have that eerie feeling of *I've lived through this before, I've seen this before*, or *I've been here before.*[2] Unlike love, déjà-vu feelings pass quickly.

Also different from love is the forgetability factor. Most of us know we've felt déjà vu, but we can't remember where or when.

Love draws us into an analysis of our own mental processes. We grow eager to comprehend our own feelings. Déjà vu does the same, albeit only briefly. Wouldn't you love to figure out why that sensation of "been here, done this" came over you when first you set foot in the temple at Karnak, attended a political rally for an obscure candidate, or played golf against a new opponent only to feel that the game had been played before—and that the bogey on the twelfth was a precognitive revelation. The feeling can be strange enough and intense enough to set you plotting episodes for *The Twilight Zone*. Perhaps you were a servant or a priest at Karnak in your past life as an Egyptian. Maybe that candidate ran for office in an alternative reality. Maybe you played that golf game in some parallel, antimatter universe.

Or maybe your brain is playing tricks on you.

WHAT IS DÉJÀ VU?

Déjà vu has been with us since the dawn of history, and since we first grunted our way out of caves, we've tried to understand it, explain it, rationalize it. It's not like any other experience. It's not a flashback, an illusion, a false memory, or a hallucination. There is no sense of depersonalization or dislocation—no feeling of being outside one's body or looking down on oneself. In fact, quite the opposite. Déjà vu is very much "here" and very much "now," just as it was, it seems, when the same thing happened before.

There's no shortage of déjà-vu accounts in literature (the Rossetti poem above among them) and no shortage of explanations. Past lives and reincarnations are popular hypotheses. Rossetti's poem serves up that notion, as does the nineteenth-century Dutch author Louis Couperus in the novel *Ecstasy:*

> Her head sank a little, and, without hearing distinctly, it seemed to her that once before she had heard this romance played so, exactly so, in some former existence ages ago, in just the same circumstances, in this very circle of people, before this very fire, the tongues of the flames shot up with the same flickerings as from the logs of ages back, and Suzette blinked with the same expression she had worn then.[3]

Some thinkers and writers have posited less supernatural explanations. Perhaps déjà vu occurs because present reality resembles some past dream or memory. Writes Ivan Goncharov in his novel *Oblomov:*

> Whether he dreams of what is going on before him now, or has lived through it before and forgotten it, the fact remains that he sees the same

> people sitting beside him again as before and hears words that have already been uttered once.[4]

Although visual theories have been proposed to explain déjà vu, vision is unnecessary for the experience. Chris Moulin at the University of Leeds studies déjà vu. Some of his research subjects are blind people who experience déjà vu just as vividly as sighted people do.[5] And, although déjà vu literally means "already seen," vision is not the only sense that can be involved. Various authors have coined a veritable déjà dictionary: *déjà entendu* (already heard), *déjà pensé* (already thought), *déjà gouté* (already tasted), *déjà revé* (already dreamed), and *déjà rencontré* (already met).

DÉJÀ AND DISEASE

Despite its commonplace nature, déjà vu can be associated with mental disorders in certain cases. In some studies, for example, schizophrenia patients actually report *fewer* déjà-vu experiences than healthy people do, but their episodes last longer and create greater feelings of unease.[6] Other researchers find a *greater* incidence of déjà vu among neuropsychiatric patients, and some evidence suggests the nature of the experience may be different.[7] People with schizophrenia tend to experience déjà vu when stressed either mentally or physically, but the content of their déjà vu, some researchers say, is no different from anyone else's.[8]

Some people who have epilepsy report a vivid sense of déjà vu preceding a seizure. The reason lies in the brain. The rapid firing of neuronal circuits in the temporal lobe, including the hippocampus, can precede both the seizure and the sense of déjà vu, perhaps explaining both. Researchers have triggered déjà vu by stimulating those brain areas with electrodes,[9] but such research may not explain garden variety déjà vu, in part because epilepsy-provoked déjà vu doesn't carry with it that same sense of disorientation—even the paranormal—that healthy people report after experiencing that been-here-before sensation.

Frequent and long-lasting déjà vu can become a pathology in itself with a name of its own, *déjà vécu:* a persistent or chronic state of déjà vu. Brain scans in such cases sometimes reveal diffuse changes in the temporal lobes, but no simple cause–effect relationship is apparent. Moulin describes three illustrative cases, all among elderly people, in which *déjà vecu* goes along with poor short-term memory and denial that the déjà vu is an illusion:

- An eighty-year-old man stopped watching television programs and reading newspapers because he believed he had seen them all before and knew their contents and outcomes.
- A seventy-year-old woman complained to her doctor that she had a bad memory. Her husband suspected she was psychic, believing that she could

predict the future because of her strong sense of having lived through events before. For example, after she heard about a terrorist attack in Bali, she claimed she knew the casualty count before the report came in.

- A sixty-five-year-old woman suffered a stroke and developed epilepsy. The first time she met a member of Moulin's research team, she kissed the scientist on the cheek and greeted her visitor like an old friend. When informed that this was a first meeting, the woman said she felt that the researcher had visited her hundreds of times before.[10]

Jamais vu ("never seen") is the opposite of déjà vu. It is the perception that a familiar object or occurrence is unknown, and it may be easier to evoke in normal brains than déjà vu. Try this: Think long and hard about the word *that*. Turn it and twist it in your mind. Write it over and over again on paper. Use it in sentences and examine its meaning from every angle. After a while, you may find yourself having no idea what the word means or even how it is spelled, although you use it regularly everyday without difficulty. "*Jamais vu* is a temporary sensation of unfamiliarity. It may include a sensation that something is new to you even though you've known it for a long time. It may include feelings of unreality, novelty, or merely a lack of fluency," says Moulin.[11]

The "word alienation" you induced when you overthought the word *that* is one example, but Moulin offers others: "[g]etting very briefly lost in a very familiar place, a loved one looking different or like a stranger, losing your way in a very familiar musical piece if you're a musician, or forgetting what pedal does what when you're driving." Moulin says 40 to 60 percent of us experience *jamais vu*, but it's reported less frequently and by fewer people than déjà vu. Ordinarily, it's no big deal, but when it becomes continuous and long lasting, it can be pathological. Neurologists describe cases in which victims of stroke or injury insist that family members are imposters. The patients may recognize facial features and even acknowledge a sense of familiarity but stand firm in their conviction that loved ones are actually strangers.

WHAT CAUSES DÉJÀ VU?

Although theories abound, no one has yet determined what's happening in your brain when you experience déjà vu. Psychologist Alan Brown has written the definitive book on the subject: *The Déjà Vu Experience: Essays in Cognitive Psychology*.[12] In his book, Brown explores several categories of possible causes for déjà vu. I've posed them as questions.

1. *Is something misfiring in my nervous system?*

   Theorists have constructed several different versions of this idea, all centering around the fact that sensory messages travel across multiple pathways from a sense organ to the brain and within the brain itself. If something goes wrong at the synapse—the juncture of two neurons and the gap between them—the travel time in one pathway might be a little longer or a little shorter than the travel time in another. This "time lapse" might lead the brain to conclude that certain information has already been received and is, therefore, remembered.

2. *Am I overgeneralizing?*

   Perhaps a few details trigger an erroneous recognition of an overall pattern—one that may seem familiar—thus prompting déjà vu.
   Brown offers an example:

   > Recall [your] first visit to a friend's home. . . . [T]he room has a layout similar to the one in your aunt's house: a sofa to the right of the love seat with a stairway to the left going up the wall, a grandfather clock against the back wall, and an Oriental rug on the floor. None of the elements in the newly entered living room is identical to one from the previous context, but the particular configuration of elements fits the same template.[13]

3. *Am I getting the sense of remembering something when I'm not actually remembering?*

   Consider ordinary remembering, without that feeling of déjà vu. You recall something. At the same time, you know you are recalling it. But what if you get that sensation of remembering without actually pulling up a memory? Suppose the *process* of remembering comes uncoupled from the *awareness* of remembering. If that happened, you might experience déjà vu—feeling as if some occurrence in the here and now happened before—although you can't recall when.

4. *Was I distracted before but I'm paying attention now?*

   This "double perception" theory goes like this: You see something once, but you are paying attention to other things, so the sight doesn't register in your conscious perception. Nonetheless, your brain stores the image away. Later, when you see the same thing and pay attention to it, you get that disconcerting feeling that you've seen it before.

For most of these ideas, theorizing abounds, while solid evidence is scant. However, Brown, a professor at Southern Methodist University, and Elizabeth Marsh, a psychologist at Duke University, actually tested the distraction hypothesis. They ran a series of experiments using students at both schools. The students tried to spot crosses superimposed on background photographs; the photographs showed places on both campuses. The students saw the photographs for only half a second. After a three-week break, the students looked at more photographs—

some new ones and some they had seen before. Between 70 and 91 percent of the students thought they definitely or probably had visited one or more sites on the other campus, although in truth they had never traveled there. Most described some sense of confusion about whether they had been there, and about half said they felt an eerie sensation of déjà vu or something like it.[14]

YOUR BRAIN ON DÉJÀ VU

Although the causes of déjà vu remain a mystery, more is known about who reports it. You are more likely to experience everyday, ordinary, garden-variety déjà vu if you travel a lot, hold liberal political views, have a good memory, or frequently exercise an active imagination. You are also vulnerable if you're tired or stressed. More money and more education make you prone to déjà vu, as does being age twenty or so. Middle-aged and elderly people hardly ever experience it, although it may turn up as part of a constellation of symptoms associated with strokes or Alzheimer's.

Is Déjà Vu in Your Genes?

No one will ever find the gene for déjà vu, but one particular gene and its inheritance patterns do intrigue molecular scientists interested in déjà vu. The gene is called *LGI1;* it lies on chromosome 10. Certain forms of the gene are associated with a type of epilepsy called autosomal dominant lateral temporal epilepsy (ADLTE). If one of your parents has that disorder, chances are (usually) 50-50 you'll have it, too. It's a mild form of epilepsy that usually appears in people in their early twenties. It involves recurrent seizures with an aura and (usually) auditory hallucinations. Though by no means a certainty, déjà vu occurs often enough during the seizure that researchers suspect a link to the gene.

People who have ADLTE have mutations (changed forms) of the *LGI1* gene, but not everyone with an *LGI1* mutation experiences seizures-cum-déjà vu.[15] Nobody knows what *LGI1* does. It may have something to do with the migration of neurons during fetal development, or it may play a part in controlling what substances get into neurons through the cell membrane. One thing's for sure: it gets around. *LGI1* operates in many different parts of the brain, guiding the manufacture of at least two different proteins.[16] Different mutations of *LGI1* produce different clinical symptoms.[17] "Thus, the gene may be expressed or responded to differently in individuals who experience déjà vu," says neurologist Edward Wild, who thinks further study of the gene may yield a "neurochemical explanation for déjà vu."[18]

Understandably, no scientists have yet captured an MRI or PET image of the brain in the midst of a déjà-vu experience. Researchers have, however, uncovered some clues that suggest where the brain's déjà-vu response may be centered. Investigators in France used stereo electroencephalograms (EEGs in three dimensions) to evaluate surgical options for twenty-four patients with epilepsy. In doing so, they applied direct electrical stimulation (harmless and sometimes therapeutic) to two parts of the brain called the perirhinal cortex (PC) and the entorhinal cortex (EC). Both lie near the hippocampus and amygdala, brain structures known to be important in forming memories. The stereo EEGs showed that stimulation of the PC often prompted memory retrieval. Stimulation of the EC often produced a feeling of déjà vu.[19]

Scientists in the United Kingdom and Australia are using hypnosis in hopes of producing déjà vu in the laboratory. The researchers, led by Akira O'Connor at the University of Leeds in the U.K., hypnotized twelve highly suggestible subjects. Under hypnosis, six of them played a game and then received the posthypnotic suggestion to forget the game; they were the posthypnotic amnesia (PHA) group. The other six did not play the game, but they were told under hypnosis that the game would seem familiar to them later on; those people formed the posthypnotic familiarity (PHF) group. Later, after playing the game for the first or second time, the subjects were interviewed and their spontaneous comments recorded and analyzed. None of the PHA participants experienced a strong sense of déjà vu when replaying the game, but five of the six PHF subjects did, although they had never actually seen the game before. "Their déjà-vu experiences were characterized by perceptual salience and confusion," O'Connor says, "very similar to déjà vu 'in the wild.'"[20]

Salience? Confusion? Wildness? That brings us full circle: Déjà vu is like love . . .

Chapter 28

Feeling a Phantom

My psychotherapist friend Dr. P. told me about one of her patients, an elderly woman we'll call Myra. Myra was five when her mother told her that her father was having an affair. She was ten when her mother left her in the care of her maternal grandmother. She was twelve when she was first told how beautiful she was. At age sixteen every boy in the high school was vying for her attention, but Myra was shy, almost reclusive.

Today Myra is a petite woman at five-feet-two and one hundred ten pounds, and still beautiful at age eighty-four. Her blue-green eyes dazzle and laugh as they mix with her infectious smile. Her hair is thick and worn in a pageboy style reminiscent of the 1950s. She has osteoporosis and its telltale sign, a dowager's hump.

She came to therapy to discuss the death of her second husband. She and Carl were married for thirty-five years. The couple had not yet celebrated their second anniversary when Myra received the diagnosis of breast cancer. Myra had always had fibrocystic breasts and lumps, but Carl didn't believe she had cancer at age fifty. He said the doctors were in collusion and trying to make money off women by removing breasts.

Myra was terrified, then mortified, and then Carl went on his "it's not cancer tirade" and she settled for terrified. Despite his objections, she had two radical mastectomies.

What was the hardest part? asks Dr. P. Myra responds,

> The hardest part was wondering what they do with the breasts once they remove them. I wondered if they just put them in a trash bag and then

they end up in a landfill or do they grind them up in a disposal. Your breasts are a sacred space; they define a big part of what it means to be a female.

It must have taken a long time to recover from two separate radical mastectomies.

It did. It still does in ways. I went to a plastic surgeon about a month after Carl died and asked if he could do anything. He gasped when he saw what the surgeons did to me. I am so disfigured. Scars crisscross every which way, and scar tissue has erupted in fissures. It is very ugly. I wish they could do something, but the doctor said that it would be very difficult since it has been so many years. He asked me, 'Why now?' And you know what I said? I told him that now that Carl is dead it is time to start living. I also told him that my breasts have kept speaking to me. He smiled and patted me on the shoulder when I said that.

What did you mean when you said your breasts speak to you?

Well, they've never stopped speaking to me. You are going to think I am crazy. I still feel them. I even can feel my nipple on my left side harden when I am chilled. There is no nipple. So anyway, I feel them and they speak to me by still being present. It was worse in the early years, but it continues. Sometimes they itch. If I don't have my form bra on I will rub my hands over my chest just to make sure they are gone. All I feel is the scar tissue. Because I had fibrocystic breasts, I was always aware of three lumps to the left of the right nipple. Since losing the breast I sometimes feel this feeling like when I had the lumps. It feels like a tingling and pressure. I know the spirit of my breasts is still alive because I still feel like I have breasts. I may just have to be satisfied to have their spirit and these reminders of what was.

WHAT IS A PHANTOM?

The sensation that Myra feels comes from a phantom. Her breasts are no longer there, but she feels them nonetheless. Many women who've had mastectomies continue to feel the missing breast, but phantoms are more often reported by amputees or accident victims who have lost all or part of a digit or limb—a finger, hand, arm, foot, or leg. Many people who have lost a limb perceive the missing body part as moving on its own or moving when willed to do so. Some feel that the limb is paralyzed or numb. In many cases, the phantom is extremely painful—thus the terms *phantom pain* or *phantom limb pain.*

Phantom pain is not the same as the pain of surgical healing or of damaged nerves that regrow abnormally in scar tissue. It's pain in what feels like the missing body part. In addition to pain, sensations of touch, temperature, pressure, itch, vibration, tingling, or "pins and needles" may be perceived in the missing body part. Nearly all patients with an amputation experience the sensation of a phantom;[1] 50 to 80 percent develop phantom pain.[2]

That pain is rooted in the reorganization of the brain's "body map" after amputation. Normally, the brain regions that process sensations from the skin—collectively designated the somatosensory cortex—are organized into specialized regions that constitute organized "maps" of the skin's surface. Certain regions receive inputs—whether pressure, heat, cold, or pain—from a particular body part. The famous Penfield map, named for its creator Wilder Penfield (1891–1976), was drawn when Penfield applied electrical stimulation to the brains of hundreds of epilepsy patients he was treating (Figure 1). He found that pinpoint stimulation of a particular brain area produced a sensation perceived in a particular body part, perhaps tongue or foot or genitals. The Penfield map shows how the perception of body parts wraps itself in a band across the top of the parietal lobe, with thumb lying above the face and the face above the tongue, like Canada, the United States, and Mexico in a neat north-south row.

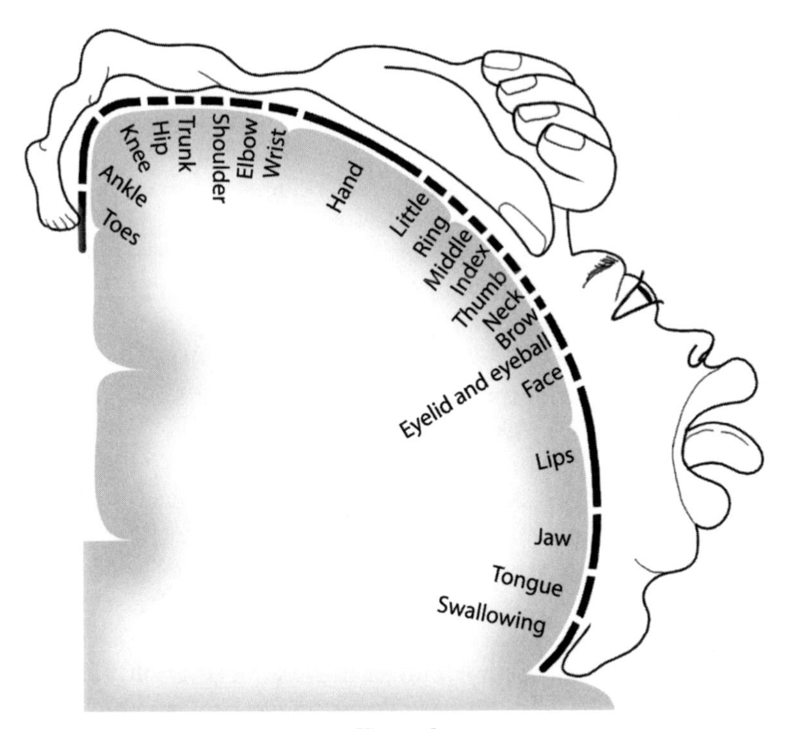

Figure 1

When a limb or other body part is lost, drastic changes occur in the map. The brain regions where input from the lost body part was once handled are taken over by nearby areas. For example, if a person has a hand amputated, the face-representing area of the cortex that lies nearby can creep into the hand-representing area. When that happens, the patient feels a referred sensation; a touch on the face feels like a touch on the hand. So just as political boundaries between nations can change on a map, the space allocated to a particular body region can change in the brain, sometimes by as much as two or three centimeters,[3] which, in geographic terms, would be something akin to Canada annexing half of Montana.

The brain's body map undoubtedly has something to do with the phantom breasts that Myra feels, and reorganization of the map must be related to phantom pain, although no one know exactly how. It's a chicken-and-egg question. Does reorganization of the brain map somehow promote the pain? Or do aberrant signals from touch nerves in the area produce both the reorganization and the pain that goes along with it? Questions persist also about the relationship of phantom pain to the posttraumatic stress disorder that for so many patients accompanies the loss of a body part. Our most emotional memories are the ones we store best, so changes in the brain's limbic system and hippocampus are implicated in phantom pain, too, although no one knows exactly how.

RESEARCHERS CHASING PHANTOMS

University of Montreal researcher Emma Duerden is looking for some of those answers. She works with the most severe cases of phantom pain. Mostly she sees patients who have suffered traumatic amputations as a result of an accident. Some of them lost their limbs in car crashes or in work-related mishaps in which limbs were caught in moving machinery. "Their lives have entirely changed due to their chronic pain," Duerden says. "They have to undergo physical rehabilitation in order to learn how to rebalance their bodies after the loss of the limb, and they often are in intense pain. Patients most times describe their pain as burning, or as if their skin is being grated. Some patients limbs feel entirely frozen and as if clenched in a vise. Duerden says,

> The pain affects their sleeping habits and, in conjunction with pain medications, their memory and attention skills can suffer. They have to take time off work or school, and in some instances they lose their social networks. . . . The pain, lack of sleep, and physical exhaustion from using a prosthesis affect every relationship they have in their lives. Some patients that I have seen were divorced after several years into their new situation in life. Needless to say, depression can be common among these individuals. What can compound these feelings is the lack of understanding of others.[4]

Duerden is one of a phalanx of bright young researchers now investigating the causes of phantom pain and experimenting with treatments. She's a doctoral candidate at the University of Montreal, but her interest in the brain's organization goes back nearly a decade, when she worked as a research assistant in a speech pathology laboratory. There she met aphasic patients whose lives had been turned upside-down by their loss of communication and physical abilities. At the same time, Duerden was learning about brain-imaging techniques, creating atlases of subcortical structures. She met patients who were having neurosurgery for Parkinson's disease and chronic pain, and she experimented with various training regimens that could help them in their recovery. For example, healthy volunteers who train over several weeks can learn to detect increasingly close-spaced points of contact on their fingertips. As a result, the area of the brain that handles touch input from the finger expands. Duerden thinks findings like those can help patients with phantom pain.

EXPERIMENTING WITH TREATMENTS

Doctors and researchers have experimented with numerous potential treatments for phantom pain. Some therapies have achieved mixed success, but none has emerged as a cure, and no factors can predict whether a therapy will work for an individual patient. Physicians often try medications first. Pain relievers such as acetaminophen (Tylenol), nonsteroidal antiinflammatories (ibuprofen and naproxen), and stronger stuff such as various opioids can sometimes help. Some benefits have been derived from antidepressant drugs, which alter levels of neurotransmitters in the brain, notably serotonin and norepinephrine.[5] Drugless approaches often include massage, hypnosis, biofeedback, acupuncture, and meditation.

Neurosurgeons have tried their hand at phantom pain relief. They've blocked peripheral nerve signals and injected painkilling agents into the epidural space of the spine. They've tried sympathectomy, a surgical procedure that cuts or destroys a cluster of nerve cell bodies in the spinal cord; rhizotomy, the cutting or blockage of nerve roots in the spine; and cordotomy, a surgical severing of pain-conducting fibers in the spinal cord.[6] Results have been mixed.

Another potential treatment is the deep brain stimulation procedure sometimes used to help patients with Parkinson's disease. To effect this treatment, a battery-operated neurostimulator—similar to a heart pacemaker and about the size of a stopwatch—is implanted in the brain. The stimulator delivers electrical impulses to specific areas of the brain, perhaps blocking the abnormal nerve signals that trigger the perception of phantom pain.

Transcutaneous electrical nerve stimulation (TENS) delivers an electrical stimulus to an area of the skin, activating one category of nerve fiber that may, in turn, inhibit the activity of pain-signaling fibers. Some theorists explain the effect as a gate control mechanism: the electrical stimulation closes the gate (through

Five Centuries of Phantoms

The phantom phenomenon is nothing new. In the mid-1500s French surgeon Ambroise Paré treated numerous patients whose limbs were amputated, mostly as a result of gangrene. "Verily it is a thing wondrous strange and prodigious," he wrote. "Patients . . . have many months after the cutting away of the Leg, grievously complained that they yet felt exceeding great pain of that leg so cut off."[7] Paré was among the first to write about phantom pain; before his innovations of ligating large vessels to staunch blood flow and applying tourniquets above the site of the amputation, few amputees survived to develop it.[8]

The philosopher René Descartes used phantoms to argue that the senses are untrustworthy, unreliable deceivers of the mind and brain:

> Many experiences little by little destroyed all the faith which I had rested in my senses; . . . I have learned from some persons whose arms and legs have been cut off, that they sometimes seemed to feel pain in the part which had been amputated, which made me think I could not be quite certain that it was a certain member which pained me, even although I felt pain in it.[9]

The Scottish physician William Porterfield (1696–1771) took a different approach. Rather than questioning his senses, he theorized that his brain was creating the perception of his own phantom, which he experienced for (apparently) some number of years after his leg was amputated:

> I sometimes still feel Pains and Itchings, as if in my Toes, Heel or Ancle, etc. tho' it be several Years since my Leg was taken off. . . . [T]ho' all our Sensations are Passions or Perceptions produced in the Mind itself, yet the Mind . . . always considers them as belonging either to the Object, the Organs, or both, but never as belonging to the Mind itself, in which they truely are. . . . Information must be carried to the Mind, and the Mind must have the same Sensation, and form the same Judgment concerning it, viz, that it is at a Distance from it, as if in the Toes, Heel or Ancle, tho' these have long ago been taken off and removed from that Place where the Mind places the Sensation.[10]

American physician Silas Weir Mitchell (1829–1914) wrote extensively about phantom sensation and pain, his knowledge coming largely from his experiences at Turner's Lane Hospital in Philadelphia. There Mitchell treated soldiers who lost limbs in the American Civil War. Mitchell also encountered phantom breast pain, and even cases of perceived erections experienced by men who lost their penises.

In his book *Injuries of Nerves and Their Consequences*, published in 1872, he wrote, "Nearly every man who loses a limb carries about with him a constant or inconsistent phantom of the missing member, a sensory ghost of that much of himself, and sometimes a most inconvenient presence, faintly felt at times, but ready to be called up to his perception by a blow, a touch, or a change of the wind."[11] Among ninety cases of amputation, Mitchell found only four in which phantom sensation was absent. All others felt the limb was still there, or as one of Mitchell's patients put it, "If . . . I should say I am more sure of the leg which ain't than the one which are, I guess I should be about correct."[12]

Mitchell published several case studies of patients who experienced phantom sensations, sometime to the detriment of daily activities:

> One of my cases attempted, when riding, to pick up his bridle with the lost hand, while he struck his horse with the other, and was reminded of his mistake by being thrown. Another person, for nearly a year, tried at every meal to pick up his fork, and was so disturbed emotionally at the result as frequently to be nauseated, or even to vomit.[13]

Despite four hundred years of reports, it was not until 1941 that phantom limb pain gained credence among most medical professionals. In that year Allan Bailey and Frederick Moersch reported on more than one hundred patients at the Mayo Clinic who either had limbs amputated at the clinic or came seeking relief from phantom pain. Most reported pain of a burning, aching, or cramping type; some felt a tight, wirelike band constricting around the phantom. Others described the sensation as crushing, twisting, grinding, tingling, or tearing. Some felt prickles in the phantom, while other said the phantom felt numb. Some felt the phantom frozen in an awkward, painful pose, while others perceived the phantom as moving of its own accord, often in painful ways.

Bailey and Moersch employed more than fifteen different treatments, but none produced more than temporary relief, and the researchers concluded that no single theory could explain phantom pain. Their best guess (for this was a time when Freud reigned supreme): obsessional neurosis.[14] So phantom pain remained a psychological aberration until the 1990s, when researchers discovered the reorganization of the somatosensory cortex after amputation.

the spinal cord leading to the brain) that phantom pain impulses would normally travel through.

Another promising approach is mirror-box therapy. The idea is to fool the brain into seeing the phantom as real, moving it at will, and experiencing (perhaps) a concomitant reduction in pain. Here's how the mirror box works. Suppose the left hand has been amputated. Watching the movements of the normal right hand in a mirror, the patient "sees" the missing left hand moving. Mirror training involves trying to perform movements with the amputated limb while viewing the movement of the normal limb. Thus, the mirror tricks the brain into seeing normal movement in the limb that isn't there, while the brain is sending signals in an attempt to elicit such movement in the phantom.

The trick seems to work in many cases. In 2007 researchers reported a trial in which twenty-two patients with phantom limb pain either practiced with a mirror, or did the same practice but with a covered mirror, or closed their eyes and imaged movement of the phantom. The patients practiced for only fifteen minutes daily; they reported their pain over four weeks. Mirror training reduced the number and duration of pain episodes and the intensity of the pain for all members of the mirror-training group. No such improvement occurred among the other patients. Nine patients from the other groups switched to mirror therapy, and eight of them noted improvement in four weeks.[15] Why? No one knows, but it's possible that visual signals might override the perception of pain. In essence, the brain might be fooled into thinking, "My limb is moving normally, so it must not hurt." A similar approach is employed in various virtual reality therapies that seek to convince the brain that the limb is still a functioning limb.

"Therapies such as mirror-box training and virtual reality are all based on the concept that the brain, once subserved with an active representation of the missing limb, will be able to harness control over the pain," Duerden says. "Largely, these theories are consistent with the idea that the brain's body maps have become distorted and that active 'use' of the phantom will increase the missing limb's representation in the brain, a sort of reverse back to the original organization." Duerden hopes to develop other training experiments to coax the brain into reorganizing its body region maps. She's planning to integrate training with neuroimaging techniques, allowing patients to view and perhaps control the representation of their brain's body maps.

Myra is not a candidate for mirror-box therapy and she won't be working with Duerden. Her phantom breasts are persistent but not physically painful. Myra lives with other phantoms that induce pains of a different sort. Says my therapist friend Dr. P.:

> For many women phantom breast pain is a haunting. This haunting urges them to look at loss, to explore grief, and to weave together the totality of all felt sensations, pain, and feelings. Myra experienced many losses

long before her breasts were removed. Her father left her, her mother left her, then one boyfriend after another left her, too. Her first husband had many affairs. She left him. Her second husband, Carl, suffered from posttraumatic stress disorder; he was never emotionally available to Myra. Nevertheless, after Carl died, Myra had a mounting feeling of anxiety. Her doctors recommended counseling. On her first session she said, 'I have always been proper. This is the first time I am not going to be proper and this is the only place I can imagine doing that. I am going to show all my anger.' Therapy is helping Myra be the person she wants to be rather than the person she felt she had to be. She uses her phantom sensations to remember all of who she is. She uses them to let go of things she cannot change. The haunting inspires her.

Chapter 29

Probabilities and the Paranormal

Rate Your Belief in the Paranormal

Using a five-point scale, where

1 = strongly disagree,
2 = disagree,
3 = uncertain,
4 = agree,
5 = strongly agree,

respond to each of the following statements:

1. The human spirit lives on after the human body dies.
2. Some people can use their minds to lift themselves or others into the air (levitation).
3. The mind or spirit can leave the body and travel (astral projection).
4. Some psychics can accurately predict the future.
5. Horoscopes predict the future.
6. The spirits of the dead can come back as ghosts.
7. Witchcraft exists, and it can be a powerful force for good or evil.
8. Some people have a lot of good luck. Some people have a lot of bad luck.
9. Some places are haunted by ghosts.

10. Some numbers are lucky. Some are unlucky.
11. The spirit can be reincarnated (reborn) into another body at another time.
12. Some mediums can call the spirits of the dead and entice those spirits to speak or materialize.
13. It is possible to be possessed by a demon.
14. Some people have the power to move objects using only their minds (telekinesis or psychokinesis).
15. Some people can read other people's minds.
16. Some psychics can use extrasensory perception to find lost objects or solve crimes.
17. Some people can communicate just with their thoughts (telepathy).
18. Some people can heal themselves or others of disease or injury using only their mental or spiritual powers.
19. Some people know what's going to happen before it happens (precognition).
20. Dreams can predict the future.

Rate your score as follows:

20 to 45: You are a diehard skeptic. You reject most, if not all, paranormal beliefs. The lower your score, the less your tolerance for what you regard as delusional thinking. Have you considered moonlighting as a professional debunker?

46 to 66: You're not a staunch believer in the paranormal, but you're not a disbeliever, either. You sit on the fence, perhaps agreeing with Hamlet: "There are more things in heaven and earth, Horatio, than are dreamt of in your philosophy."

67 to 100: You are a believer in paranormal phenomena. Scores between 67 and 80 show that you question sometimes, but not often. Above 80, and you're a believer without reservation. The world of the paranormal is very real for you. Doubts are out of the question.

Nobel Laureate Charles Richet died in 1935, but were he alive today, I'm guessing he would score somewhere around 65 on my paranormal propensity quiz. The author of *Thirty Years of Psychical Research,*[1] Richet coined the term *métapsychique* from the Greek meaning "beyond the soul." The term has been translated into English as metapsychics, and Richet spent a lifetime trying to get other scientists to take seriously the clairvoyance, telekinesis, and telepathy that so fascinated him.

METAPSYCHICS MEETS SCIENCE

Richet became a respected scientist not because of his psychical research, but in spite of it. The Frenchman was a true polymath. He followed in the footsteps of Pasteur and Lavoisier, studying lactic fermentation and oxidative metabolism. He did groundbreaking work on the chemical action of gastric enzymes. He investigated the effects of anesthetics on reflexes. He studied the resistance of bacteria to metallic poisons and the mechanisms by which animals regulate their body temperature, including panting in dogs. He measured oxygen conservation in diving ducks and was the first to postulate the action of neurotransmitters. Not to limit his scientific endeavors to physiology, he designed a monoplane that flew before the Wright Brothers did. Richet won the Nobel Prize in 1913 for his discovery of anaphylaxis, the extreme allergic reaction to a foreign protein that results from previous exposure to it. He did not confine his erudition to science. He wrote many poems, plays, and novels, some of which achieved critical acclaim in literary and dramatic circles.[2]

But at the end of his life, as in his early years, Richet was fascinated above all else with metapsychics. As a young man, he became proficient in hypnotism and used it to effect temporary "cures" for various maladies suffered by some of his friends. While still serving his medical residency, he became fascinated with clairvoyance, and he published an article on the subject in a nonscientific journal. In his article, he applied the mathematics of probability to experiments in which a person in one room guesses which playing card has been turned over by a person in another room. The better-than-expected findings of such experiments could not be attributed to chance, Richet concluded, and neither could instances of precognition, in which future events had been predicted by individuals gifted with extrasensory perception.

In the final decade of his life, Richet devoted himself solely to what his peers derided as spiritualism and the occult, but Richet rejected any notion that supernatural forces were at work. "The terms 'supernatural' and 'supernormal' . . . are . . . both inadmissible, for there can be nothing in the universe but the natural and the normal," Richet wrote.[3] He asserted that the brain is the source of extrasensory perception in all its forms, making telepathy, telekinesis, and even the materialization of spirits nothing more—and nothing less!—than a sixth sense. While Richet had little doubt that the human mind had such capabilities, he attributed them to natural laws, laws yet to be elucidated through the methods of scientific inquiry. "The facts of metapsychics are neither more nor less mysterious than the phenomena of electricity, of fertilization, and of heat. They are not so usual; that is the whole difference. But it is absurd to decline to study them because they are unusual," he wrote.[4]

A Short Course on Probability

One of the problems with the "sixth sense" is that, if it exists at all, it's not a big thing. It's a small, subtle phenomenon, and numbers get in the way. To look at the problem in a simple fashion, consider tossing coins. With one coin, the probability of heads or tails is easy to figure out: ½. The chance of a head exactly equals the chance of a tail. What is the chance of tossing five heads in a row? Since the probability of a head is the same with every toss, the correct answer is $(\frac{1}{2})^5$ or $\frac{1}{2} \times \frac{1}{2} \times \frac{1}{2} \times \frac{1}{2} \times \frac{1}{2}$, or 1/32. So, on average, in every series of 32 tosses of five, five heads will come up once. But you can perform 32 tosses of five each and never get five heads. Or you might get all heads several times. The only way to get a number that comes out very, very close to the mathematical probability of 1/32 is to toss the coins perhaps 32,000,000 times. Then you'll get 1,000,000 five-head sequences, or maybe you'll get 1,000,003, or 999,997. Chance operates even in big numbers.

Now, let's say you sand one side of the coin so that you dramatically alter its tossing properties. Now, the chance of a head is not ½ but ⅓. The probability of five heads in a row is now $(\frac{1}{3})^5$ or 1/243. That's a big change, or what statisticians call a large effect size. If you tossed a series of five coins for 500 trials, you'd nearly always notice a difference between 1/32 and 1/243, something on the order of 15 versus 2 in that many tosses.

But let's say that you changed the tossing properties of the coin only very slightly, reducing the probability of a head from ½ to 1/2.01. Now the probability of five in a row drops from 1 in 32 to 1 in 32.8. Over many, many trials, you might be able to pick up that difference and find a *statistically significant* factor operating over and above chance. But it would take thousands, maybe millions. In a small number such as 500, the random fluctuations of chance would probably mask that subtle change in probabilities.

Thus, a large effect size is easily found in a small number of trials. But if the effect size is small, it takes very big numbers to find it. In statistical terms, the power of the test is too low, and low power may explain why senses like ESP or telepathy—if they exist—are so difficult to demonstrate experimentally. The experiments simply aren't powerful enough to find them.

METAPSYCHIC RESEARCH TODAY

Richet, were he alive today, would be dismayed to find that his plea for more attention to metapsychic research has largely been ignored. Scientific studies on telekinesis and telepathy are scarce, and those investigators who conduct them often operate on the fringes of academe. Nevertheless, some efforts have been made, but few of them have yielded evidence strong enough to change the minds of those who scored high or low on my paranormal quiz.

Some studies attribute the perception of psychic anomalies to suggestibility on the part of observers. In one study, British researchers staged a fake séance in which an actor said that a table levitated. Later, one third of the participants in the séance reported that the table had risen, although in truth it had not. The reports of observed telekinesis ran higher among believers in the paranormal than among self-identified skeptics.[5] In another experiment, the same researchers showed participants a videotape of a fake psychic placing a bent key on a table. In some recorded segments, subjects heard the psychic declare that the key was continuing to bend; in other recorded segments, no such observation was stated. Those who heard the suggestions were more likely to report that the key continued to bend—and that result was statistically significant (see sidebar, A Short Course on Probability). What's more, those observers who reported that the key continued to bend were significantly more confident of their recall than were those who reported no bending. Those who "saw" the continued bending were also less likely to recall that the fake psychic had said anything about the bending key.[6]

Other researchers have used brain-wave recordings and brain-imaging technologies to investigate metapsychic occurrences and beliefs. Scientists at Harvard studied sixteen pairs of emotionally connected individuals (five couples, four close roommate/friend pairs, two identical twin pairs, one mother–son pair, and one pair of sisters). One member of each pair was classed as the "receiver," and fMRI images of his or her brain were captured while the other member of the pair, the "sender," sat in another room, viewing a photograph. The sender attempted by whatever means possible to communicate the photograph to the receiver. Inside the fMRI scanner, the receiver saw two photographs and pressed a button to indicate which one the sender was transmitting. The photographs were varied in their emotional content and in the order of their presentation.

Out of 3,687 recorded responses, the receivers correctly guessed the sender's photograph 1,842 times (50.0 percent). None of the receivers did any better than chance in guessing the correct photograph. Although the emotional content of the photographs had the expected effect on patterns of brain activation, no differences showed up when transmitted or nontransmitted images were compared or selected. "These findings are the strongest evidence yet obtained against the existence of paranormal mental phenomena," the investigators concluded.[7]

While only a little research is being done on paranormal phenomena per se, a great deal more is being conducted on people who believe in such phenomena. Some researchers have hypothesized that believers and disbelievers differ in some measure of cognitive competence or critical thinking, but investigations

The Other Side of the Coin!

Readers who scored above 70 on the quiz may want to invest in Georges Charpak and Henri Broch's book *Debunked! ESP, Telekinesis, and Other Pseudoscience* (Baltimore: Johns Hopkins University Press, 2002). A Frenchman like Richet, Charpak won a Nobel Prize in physics in 1992 for the invention and development of particle detectors. Unlike Richet, Charpak gives no credence to anything remotely connected to metapsychics. The book reveals how, for pleasure or profit, fakers can amaze observers with feats of telepathy, telekinesis, and precognition. It's all simple science, Charpak and Broch argue—and in that they agree with Richet.

have produced mixed results. Tests of academic achievement can't predict belief status, nor can concrete variables such as grade-point average or years of formal education. However, several studies have found that believers are more likely to underestimate the probability of a chance occurrence and less accepting of coincidence as a logical explanation than are nonbelievers. Some evidence suggests also that believers perform worse on tests of syllogistic reasoning than do skeptics.[8] (Examples of a syllogism: Is this series logical? All ducks are birds. All mergansers are ducks. Therefore, all mergansers are birds. Is this one? All ducks are birds. All mergansers are birds. Therefore, all mergansers are ducks.)

Under the auspices of the KEY Institute for Brain-Mind Research in Zurich, Switzerland, an international team of researchers investigated whether believers in the paranormal are more right-brain dominant than nonbelievers. The researchers recorded resting EEGs for strong believers and disbelievers. Compared to disbelievers, believers showed more right-side sources of certain types of brain waves and fewer left-right differences in other types, a finding that the researchers interpreted as evidence of greater right-brain activation in believers. The right hemisphere, they noted, is the seat of "coarse rather than focused semantic processing [which] may favor the emergence of 'loose' and 'uncommon' associations."[9] Other researchers are more charitable in their interpretation of the right-brain dominance findings. They argue that right hemisphere overactivation may explain why believers are better able to find meaningful connections between experiences and events, resulting in what the scientists termed "individual differences in guessing behavior."[10]

I suspect Richet would object to the reduction of his beloved metapsychics to idiosyncratic guessing. "Everything of which we are ignorant appears improbable, but the improbabilities of today are the elementary truths of tomorrow," Richet wrote.[11] Only a psychic can predict where paranormal research will take us in the next century. Or maybe not.

Chapter 30

Time

Body Clock

What have I been missing?
While staring at artificial time
Modern, civilized and disarming
Buzzers and beepers, alarms and reminders
Training me to ignore the subtle

Clock face down, cell phone off
Computer off, TV off, light switch off
No calendars, weather reports, GPS data
Stop screaming through my days
Release my nights from bondage

Wonder
Without plug-ins
Sighs of knowing
Without utilities
Trusting self and surroundings

I will feel again, in my blood
The sunrise
The moonrise
The Milky Way
Sundogs and clouds

I will feel again, in my bones
The smell of spring
The taste of fall
The touch of winter
The passion of summer

I will feel again, in my soul
Within the cells of my being
What I have missed
And the reunion with life
Will be joyous

—Kim McKee

The poet Kim McKee is right. If we unplug our alarm clocks, drain our cell phones of their charge, and deactivate our beepers and buzzers, we find joy in reuniting with human experience. Try this. For one week (if you can stand it and you have time away from work or school), go to bed when you are sleepy. Rise when you wake naturally and feel rested. Eat when you are hungry. Drink when you are thirsty. Ignore the clock. Don't answer the phone. Turn off its ringer if you can. Put your cell phone, Blackberry, and Bluetooth in a drawer. Put your laptop in the closet and put a sheet over your desktop computer. Turn clocks toward the wall. Turn off your electric lights.

Live in tune with life's natural rhythm and feel, if only for once in your life, what it means to be in touch with your time sense.

Time sense? Yes, you have one, and I argue that it's as powerful as vision, hearing, touch, or the chemical senses in shaping our lives. While we've been busy developing technology to replace or overpower it, we've subverted our time sense less than we might believe. We still must sleep, though we take pills to prevent it. We still lie awake, when we wish we wouldn't—but then we take pills for that, too. Some people argue that we would do well to abolish the body's natural sense of time. I argue the opposite. In these days of all-night supermarkets and twenty-four-hour banking, I believe the sense of time is the most important brain sense of all.

HOW THE BODY AND THE BRAIN KEEP TIME

The body and the brain have several built-in, automatic, self-regulating systems for measuring and keeping time. One such system is the organic "stopwatch" that lets the brain judge (more or less accurately) when five minutes have passed. Other timekeeping systems regulate monthly changes (for example, the menstrual cycle in females) and seasonal cycles (which can cause some people to feel depressed in the winter months, when light levels are low).

The most important clock for timing sleep and waking is the twenty-four-hour, circadian (meaning "about one day") clock in the brain. The circadian clock is centered in the suprachiasmatic nucleus (SCN). That cluster of cells—smaller than a pencil eraser—is located in the hypothalamus of the brain, just above where the optic nerves cross. When the SCN is absent or damaged, animals lose their daily rhythms. If they receive "transplants" of SCN cells, they regain their daily periodicity.

Signals from the SCN travel to several brain regions, including the pineal gland. The pineal secretes the sleep hormone melatonin. During the day, the SCN blocks melatonin release. In the evening, the blocking action subsides, and the pineal releases melatonin, usually about two hours before a person's habitual bedtime. Melatonin prepares the body for sleep. It induces a feeling of drowsiness and trips the "sleep switch" in another area of the brain.

The SCN uses chemical messengers and nerve impulses to regulate more than sleeping and waking. It "schedules" rises in blood pressure, heart rate, blood sugar, and the stress hormone cortisol to coincide with the beginning of the day. It also regulates the body's daily temperature cycle. Body temperature rises before waking, peaks in the afternoon or early evening, and falls to its lowest between 2 and 5 in the morning. Some of the SCN's daily cycling triggers illness. Its action may be the reason that heart attacks occur more frequently in the morning[1] and asthma attacks often come at night.[2]

The neurons of the SCN do something that no other neurons can do. On their own—even growing in a laboratory dish—they create cycles of chemical and electrical activity that correspond to a twenty-four-hour day. Genes control their timed action. A few of the genes that regulate cells of the SCN have been studied. In mammals, some of the best known are called *clock, bmal, period,* and *cryptochrome.* Those genes switch on and off in rhythmic cycles, giving the neurons of the SCN their timekeeping ability.

One such rhythmic cycle is produced by the interaction of the *period* and *cryptochrome* genes. They cause the cells to make proteins, nicknamed "per" and "cry," that are self-regulating. When the levels of per and cry become high enough, they "switch off" the action of *period* and *cryptochrome*, the same genes that cause their production. *Period* and *cryptochrome* are, in turn, switched on by the action of *clock* and *bmal.*[3] Thus, in a repeating daily cycle, genes switch on and off, creating a twenty-four-hour cycle of protein production. In this way, a molecular clock drives a cellular clock that, in turn, drives an entire organism.

The circadian clock of the SCN needs no external cues to make it run. It works automatically. It sticks to a twenty-four-hour day despite changes in periods of light and dark. Volunteers who stay in continuous light or dark for many days nevertheless maintain a twenty-four-hour cycle of sleeping and waking. Even under the influence of large amounts of caffeine, which many coffee drinkers ex-

The Bowel's Body Clock

Virtually all organs in the body have their own "clocks" and operate in concert with them. Heart rate, metabolism, digestion, breathing—all cycle daily.

The need to eliminate solid wastes, for example, is suppressed at night. It starts up again in the morning, under the direction of the SCN. The communication between the SCN and body organs depends on chemical messengers. No one knows for sure all the chemicals that keep body actions such as bowel movements on schedule, but one likely candidate for at least part of the job is the small protein prokineticin 2 (PK2).[4] It is not a part of the clock itself, but it communicates clock time to other parts of the brain and body.[5] One of its jobs is to regulate the daily rhythms of movements in the stomach and intestines. That's why bowel movements are regular, morning events and why shiftworkers and jet travelers often suffer from constipation.

perience, the SCN stays accurate to within 1 percent.[6] That 1 percent is a small, but important, difference. It gives the SCN system enough flexibility that changes in the light-dark cycle of the environment can "reset" the clock.

RESETTING THE SCN

Fine-tuning of the SCN happens every day, so that the internal clock stays in synchrony with the earth's rotation and with the changing seasons. The reset relies on nerve impulses that travel from the eyes to the brain. Extremely important in this process are the retinal ganglionic cells (RGCs) in the eyes. There are only about two thousand such cells, compared to millions of the visual cells, rods, and cones. RCGs are actually extremely *in*sensitive to light, but when they are activated, they produce a big signal that goes directly to the brain.[7] The signal results from a chemical change in the pigment melanopsin, which is a rarer pigment than other light-sensitive pigments in the eyes. Impulses from RGCs travel to the SCN, where they cause the body clock to reset.[8]

For several years, scientists thought that melanopsin was the only system that reset the circadian system, but recent evidence suggests that the circadian clock resets itself even when the melanopsin system is not operating. Researchers in Oregon showed that RGCs receive light information from rods and cones, the ordinary cells of vision. Light triggers impulses in RGCs when rods and cones release certain neurotransmitters, some of the same ones that operate in the visual system of the brain. How RGCs combine the different inputs to reset the circadian clock is not known.[9]

SLEEPING AND WAKING

The drive to sleep cannot be overcome, despite our best efforts. The brain induces it, and—happily, I'd say—we are powerless to resist it, no matter how alluring our technologies. Certain parts of the brain trigger and maintain wakefulness. Other parts put us to sleep and keep us that way. The brain's "sleep switch" is the ventrolateral preoptic (VLPO) nucleus. This grape-sized cluster of cells lies in the front of the hypothalamus, not far from the SCN. "The VLPO turns out the lights in the brain and lets it go to sleep," says Clifford Saper, a neurologist at Harvard Medical School.[10]

Like all neurons, the nerve cells of the VLPO do their job chemically. One of the most important inhibitory neurotransmitters is gamma-aminobutyric acid (GABA). The nerve cells of the VLPO release GABA. Molecules of GABA bind to neurons in another part of the hypothalamus, the tuberomammillary nucleus (TMN). When the TMN is active, the neurotransmitter histamine in the TMN promotes wakefulness. GABA from the VLPO shuts that system down. Without signals from the TMN saying, "Stay awake," sleep begins. The VLPO has no GABA receptors, so the inhibitory neurotransmitter doesn't affect it.[11]

The brain's main "stay-awake" centers are located in the hypothalamus and the brainstem. A complex cocktail of neurotransmitters regulates them. The neurotransmitters include acetylcholine, norepinephrine, serotonin, and histamine. Some scientists think another neurotransmitter, hypocretin (also called orexin), is the main chemical that pushes the "awake button" in the hypothalamus.[12] Brain cells that produce hypocretin are normally active during waking and inactive during sleep. Without hypocretin, we'd stay asleep all the time. When it's produced, however, we are awake.[13] Another awake chemical is corticotropin-releasing hormone (CRH). It promotes the awake state and may be the reason we wake naturally, without the help of an alarm clock.[14]

WHEN THE TIME SENSE FAILS

Shifts in the body's daily rhythms can disrupt work and play. One disorder of the circadian clock is delayed sleep phase syndrome (DSPS). People with DSPS can't fall asleep as early as their jobs or social lives demand, and they have trouble rising as early as they need to. They toss and turn for hours trying to fall asleep, but sleep won't come until the early morning hours. Once asleep, the person with DSPS sleeps normally, but rising when the alarm clock rings is difficult.

Advanced sleep phase syndrome (ASPS) is the opposite of DSPS. The individual with ASPS feels sleepy earlier than most people do, typically between 7

Does Time Slow Down During a Crisis?

Once, while driving on an icy road in Colorado, my car spun through a full 360-degree circle on a hill with traffic coming in both directions. At the time, the spin seemed to occur in slow motion. Minutes passed, my brain told me, during that rotation, although in truth only seconds went by. Did my time sense truly slow during that traumatic event? Researchers at the California Institute of Technology say no. To reach that conclusion, the investigators suspended twenty brave volunteers from a tether and timed them as they dropped thirty-one meters (more than one hundred feet) into a safety net. The volunteers experienced no change in their perception of time but, after the fall, they estimated their drop time as 36 percent longer than it actually was. It's the richer encoding of the memory of the event that distorts the brain's time sense, the researchers say, perhaps because the amygdala—one of the brain's emotion-processing centers—plays a major role in forming memories of emotionally charged events.[15]

and 9 p.m. Early morning waking may occur between 2 and 5 a.m. Sleep in between is normal. ASPS is much rarer than DSPS. It occurs most often in the elderly. The disorder may result from a reduced capacity to respond to daily cues that reset the SCN. One form of ASPS is inherited from one parent. The gene for it is located on chromosome 2.[16]

Another circadian disorder is described simply as irregular sleep–wake patterns. Sleeping and waking occur at unpredictable times of the day and night, in episodes of no more than four hours. This disorder is rare. It is usually seen only in people with long-term illnesses who require months of bed rest or those with brain injuries or diseases that damage the SCN.

A fourth circadian disorder is the non-twenty-four-hour sleep-wake disorder. It is extremely rare, affecting only those few individuals whose body clocks do not operate on a twenty-four-hour cycle. Their body clocks are said to be "free-running." They typically delay sleep and wake times by one or two hours daily. People who are totally blind are most likely to develop this disorder, affecting perhaps as many as three in every four.[17]

Since everyone has occasional problems getting to sleep or staying asleep, disorders in the sleep cycle are diagnosed only if the problem persists for several months. Sleep interventions are used to treat the condition, the goal being to reset the SCN. Chronotherapy, a scheduled exposure to bright lights at certain times, is one such intervention. It uses light to reset the SCN and thus change sleeping and waking times. Doses of the sleep chemical melatonin—a hormone normally produced at night and in the dark—can also help in some cases.[18]

In the end, perhaps all of us are "phase disrupted." We ignore our body clocks or attempt to destroy them. We stay up too late, get up too early, try to pack our days and nights with activities on the weekdays, only to crash and burn on the weekends. Fighting the brain's natural time sense robs us of our youth. It makes us old before our time. Resist that. Follow the poet's advice. Feel again, in the cells of your being, your brain's time sense—indeed, all your senses. The reunion will be joyous.

Appendix: The Brain and the Nervous System—A Primer

> The point for me isn't to develop a map of the brain, but to understand how it works, how the different regions coordinate their activity together, how the simple firing of neurons and shuttling around of neurotransmitters leads to thoughts, laughter, feelings of profound joy and sadness, and how all these, in turn, can lead us to create lasting, meaningful works of art. These are the functions of the mind, and knowing where they occur doesn't interest me unless the where can tell us something about how and why. An assumption of cognitive neuroscience is that it can.
>
> —Daniel J. Levitin[1]

You may have turned right past this newspaper headline: "International Team Creates First High-Res Map of the Human Cerebral Cortex." You may have been interested in sports or politics or the weather that day, and that story didn't seem to have much to do with your everyday life. But in truth, that map captures everything about life as you experience it. The cerebral cortex is the outer layer of the brain, and it's responsible for just about everything you think, feel, say, and know. The map mentioned in that headline shows how neurons in the cortex communicate in highly complicated networks.[2]

This appendix does not trace all these connections; rather, it summarizes some of the basics of the brain's structure and function. It describes how the brain maintains its two-way communication with all parts of the body. It also provides

Brain-Imaging Techniques

Scientists have several ways of looking at the brain, but ordinary x-rays aren't of much use. They can't reveal depth, so they don't show much detail in the overlapping, soft structures of the brain. The computed tomography (CT) scanner, introduced in the 1970s, was the first technology to open the door to brain-imaging research.

As the CT scanner rotates, measured amounts of x-rays travel along a narrow beam. As the beam passes through, the brain absorbs more or less of the radiation depending on the density of the tissue. Detectors convert the rays that pass through into electronic signals that are transmitted to a computer. The computer calculates the difference between what went in and what came out. From this comparison, it draws pictures of thin "slices" of the brain. It then puts the slices together into a three-dimensional image and projects it on a television screen.

Another way to look inside the skull is magnetic resonance imaging (MRI). An MRI scanner is a tube that is really nothing more than a powerful magnet. The magnet causes the hydrogen atoms in the brain's water molecules to line up in one direction. When the atoms flip back to their original state, they emit a weak radio signal. The speed of the signal depends on how dense the tissue is. From the signals, a computer can create an image. As far as we know, MRI is perfectly safe. There seem to be no health hazards associated with the magnetism it uses.

A related procedure is functional MRI (fMRI). It shows not only structures but also the organ in action. Suppose a part of the brain is at work on a problem. More blood flows to that area. The increased oxygen changes the radio signal from that area. Thus, fMRI can show which part of the brain does a certain job.

Positron emission tomography (PET), introduced in the 1980s, can do that, too. The subject is injected with a radioactive isotope. (Isotopes are atoms of an element that differ in their number of neutrons.) As it travels through the body, the isotope emits positively charged electrons, or positrons, which collide with negatively charged electrons in the body. When the particles destroy one another, they release gamma rays. PET equipment detects the gamma rays, and a computer turns the data into a colored map of "where the action is."

Another useful brain-imaging technique is magnetoencephalography (MEG). It records magnetic fields produced by the living brain without any interference from the skull or scalp tissue. These fields are extremely small, but MEG can determine their location to within two to three millimeters in less than one one-thousandth of a second.

Another promising new technology is diffusion MRI technology, a non-invasive scanning technique that estimates how nerve fibers "hook up" in the brain based on different rates of the diffusion of water molecules. One form of this method, diffusion spectrum imaging (DSI), can capture the angles of multiple fibers crossing in a single location. The technique has been used to map the trajectories of millions of neural fibers that course through the hills and valleys of the brain's surface.[3]

some diagrams to help interested readers locate various structures and regions in the brain that are mentioned in some of the previous chapters.

BRAIN STRUCTURE

Your brain contains about 100 billion nerve cells, or neurons.[4] A fissure runs down the middle of the brain, from front to back, dividing it into two hemispheres, the right and the left. The left hemisphere contains nearly 200 million more neurons than the right.[5] The two hemispheres control opposite sides of the body, so when you lift your left leg, it's your right brain that sends the command. Between the hemispheres lies the corpus callosum. This band of about 300 million nerve fibers carries messages between the two hemispheres and coordinates conscious thought, perception, and action.[6]

The hemispheres are made of a mixture of gray matter—which is actually pink in the living brain—and white matter. The gray matter is composed primarily of the bodies of neurons. It's found mostly in the brain's outer layer, the cerebral cortex, which is only about one quarter of an inch (six millimeters) thick.[7] The white matter lies beneath the cortex. It's made mostly of axons, which are thin projections from the cell bodies of neurons. Axons carry messages away from the bodies of neurons, so they are important when neurons "communicate" with one another. The fibers of the corpus callosum are the white matter of axons. They are white because the insulating protein myelin surrounds them.

Most of the brain's cells are not neurons. They are glial cells. Somewhere between one and five *trillion* glial cells protect and support the neurons of the two hemispheres.[8] The different types of glial cells include the following:

- Oligodendrocytes, which form protective sheaths around the axons of neurons
- Satellite cells, which surround the cell bodies of neurons
- Neuroglia, which guide the path of growing neurons during fetal development

Spotlight on Glial Cells

Once considered supporting actors who never got starring roles, glial cells are now coming into the research limelight. And, it turns out, while the leading actors, the sensory neurons, often get all the credit, it's the humble glial cells that are, in fact, delivering a star performance. Researchers at Rockefeller University are working with sensory organs in the humble nematode, *C. elegans* (see Chapter 5). Those organs contain both glial cells and neurons. To see what the glial cells do, the scientists removed them and observed the outcomes. The loss of glia affected the structure of the sensory neurons. They shriveled in size and lost their treelike branches. They also lost their ability to respond to stimuli. Normally, nematodes find a spot that is neither too hot nor too cold, but the glia-less animals just kept crawling toward warmer and warmer regions. They couldn't discriminate odors, either. Looking for a molecular explanation for those changes, the researchers found a protein, FIG-1 that is manufactured in the glial cells of the nematode's sense organ. When it's present, sensory neurons form properly and work correctly. FIG-1 resembles a human protein called thrombospondin, which suggests glial cells may play some of the same roles in humans as in nematodes.[9]

Another important type of glial cells is the microglia. These cells constantly extend microscopic "arms" into surrounding brain tissue where they clean out dead cells and check for damage. They spring into action—manufacturing disease-fighting chemicals—when the brain is injured or infected. Also important are the astrocytes, or astroglia. Scientists once believed that astroglia merely provide nutrients, support, and insulation for neurons, but recent research suggests they do much more. They "instruct" new, unspecialized brain cells, "teaching" them what kind of mature neuron to become. They also help regulate the transmission of nerve impulses and the formation of connections between brain cells.[10]

BRAIN REGIONS

The brain has three main regions: the hindbrain, midbrain, and forebrain (Figure 1). Each has its own specialized structures.

The hindbrain includes the upper part of the spinal cord, the lower part of the brainstem, and the cerebellum. The spinal cord is a large bundle of nerve

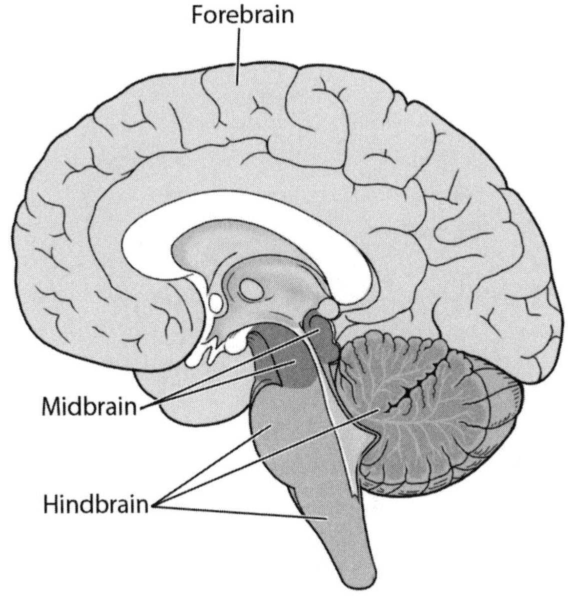

Figure 1

fibers enclosed within the vertebrae of the backbone. It carries messages between the brain and the rest of the body. The brainstem regulates heartbeat, breathing, blood pressure, and reflexes such as coughing, sneezing, swallowing, and hiccupping. Just above the brainstem lies the cerebellum. It coordinates muscle actions and maintains balance. It controls muscle tone and posture. The memories of learned activities that become automatic, such as riding a bicycle or driving a car, are stored in the cerebellum.

The uppermost part of the brainstem is the midbrain. It controls some reflex (automatic) actions. The midbrain also initiates voluntary movements of the eyes. An injury of the midbrain brings about unconsciousness or coma.

Toward the front and top of the head lies the forebrain, which includes the "thinking" brain, or cerebrum. The cerebrum has a thin covering called the cerebral cortex. This part of the brain looks bumpy and wrinkled. Its "hills" are called gyri; the "valleys" are sulci. The extensive wrinkling and folding serve a purpose. They increase the brain's surface area while packing more neurons in less space. The cerebral cortex contains two thirds of the 100 billion neurons in the brain and nearly three quarters of its 100 trillion neuronal connections.[11] In this thin layer lie the neurons that process sensory perception, voluntary movement, conscious thought, purpose, and personality (Figure 2).

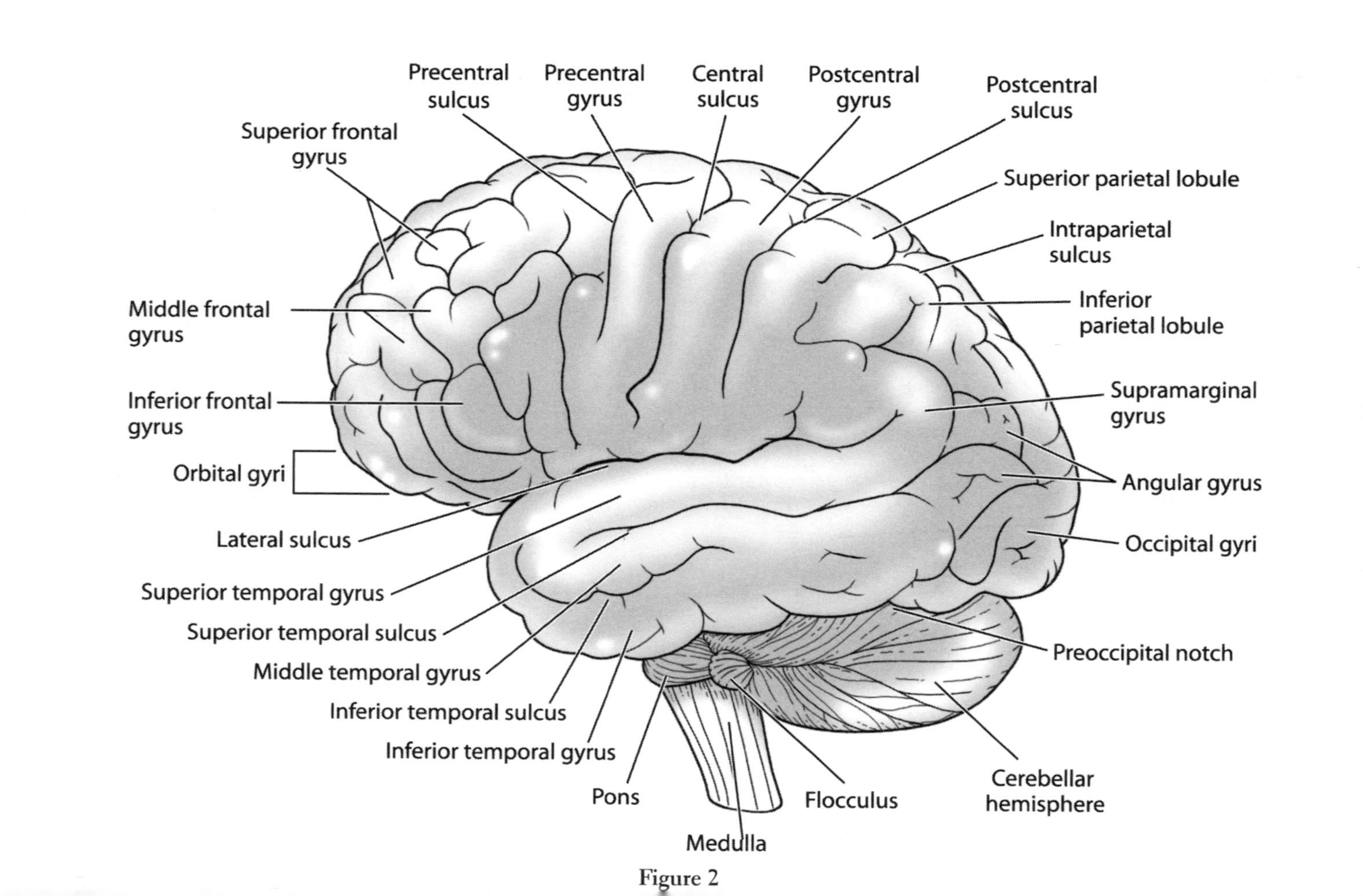
Precentral sulcus
Precentral gyrus
Central sulcus
Postcentral gyrus
Postcentral sulcus
Superior frontal gyrus
Superior parietal lobule
Intraparietal sulcus
Middle frontal gyrus
Inferior parietal lobule
Inferior frontal gyrus
Supramarginal gyrus
Orbital gyri
Angular gyrus
Lateral sulcus
Occipital gyri
Superior temporal gyrus
Superior temporal sulcus
Middle temporal gyrus
Preoccipital notch
Inferior temporal sulcus
Inferior temporal gyrus
Pons
Flocculus
Cerebellar hemisphere
Medulla

Figure 2

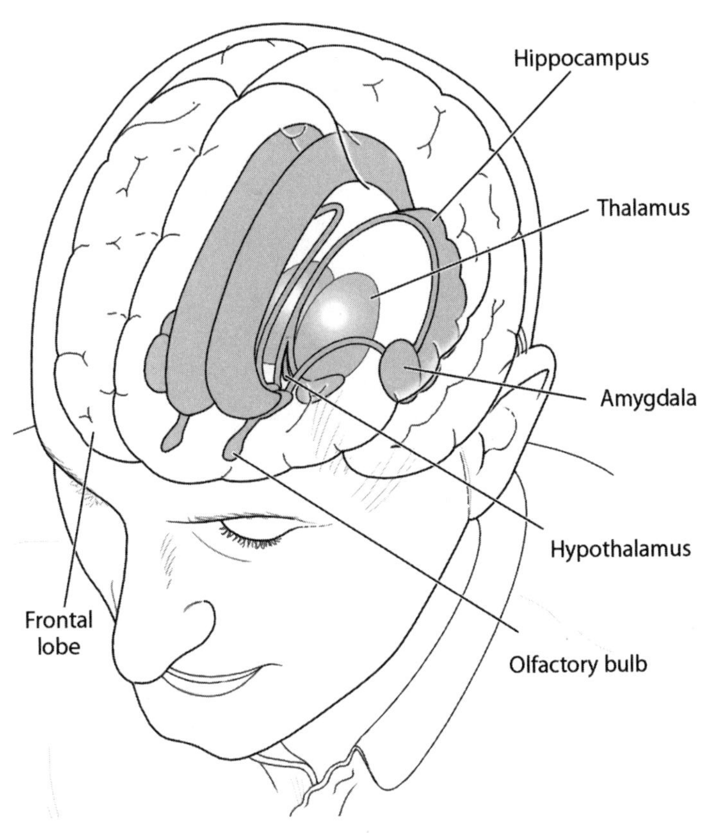

Figure 3

In the forebrain, under the cerebrum, lies a group of structures collectively called the limbic system. The limbic system includes the hypothalamus, thalamus, hippocampus, amygdala, and other structures. These structures govern emotions ranging from love and compassion to violence and aggression. They also play various roles in motivation, behavior, and some involuntary functions. The hypothalamus, for example, controls thirst, blood sugar levels, hunger, and body temperature. Near the hypothalamus lies the thalamus. It acts as "central dispatch," sorting through the messages that come into the brain via the nerves of the spinal cord and sending them to the appropriate higher brain centers for action (Figure 3).

LOBES OF THE CORTEX

The cortex has four lobes (Figure 4):

- Frontal, under the forehead toward the top of the head
- Parietal, at the crown of the head

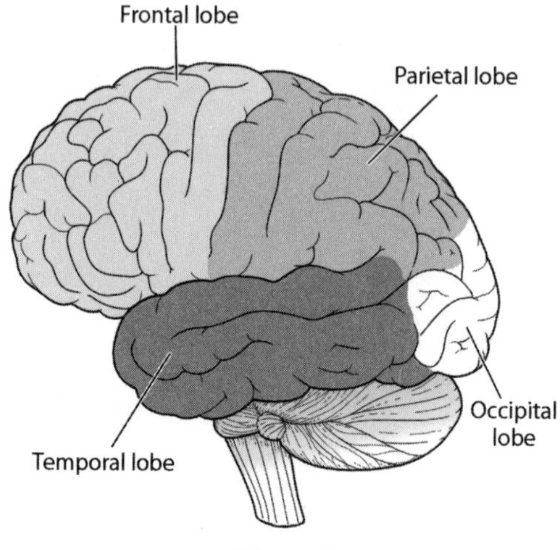

Figure 4

- Occipital, at the back of the head just above the cerebellum
- Temporal, along the side of the head behind the eyes and above the ears

The frontal lobe is the largest. It's where most planning, decision making, and purposeful behavior begin. People with damaged frontal lobes can't adapt to new situations or understand complex ideas. The frontal lobe connects with the seat of emotions, the limbic system. Much of your awareness of danger comes from your frontal lobe as it processes signals from the brain's fear centers deep in the limbic system. At the back of the frontal lobe lies the brain's motor area, where voluntary movement is controlled. Broca's area in the frontal lobe transforms thoughts into words.

In the parietal lobes the body processes the sensory information it obtains from touch. Neurons there process information about the position of the body in space, muscles, contact, and pressure. Neurons in the parietal lobe also integrate movements. The motor cortex is a thin strip that lies at the rear of the frontal lobe. It handles voluntary movements. The more a body part is used or the more sensory neurons it contains, the more space it gets in the sensorimotor cortex. For example, although your back is larger than your hand, it makes fewer complex movements and is less sensitive, so it gets less space in the parietal lobe than your hand and fingers, which are sensitive to touch and move in myriad ways (Figure 5).

The occipital lobes lie at the back of the head. Because they handle vision, they are often called the visual cortex. Information goes from the eyes through a relay station (the lateral geniculate nucleus) to the visual cortex. Damage to the occipital lobes can result in blindness even if the rest of the visual system is normal. The figure shows some additional detail on the specialized regions within the visual system of the occipital lobe (Figure 6). (See Part Four for more information

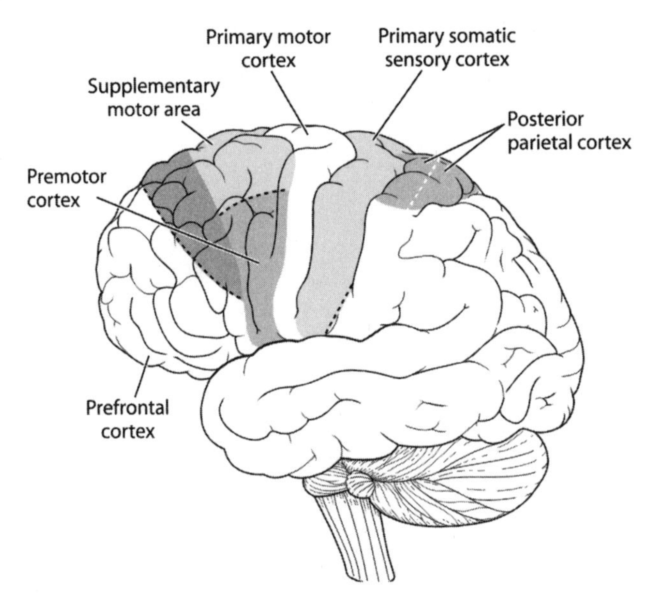

Figure 5

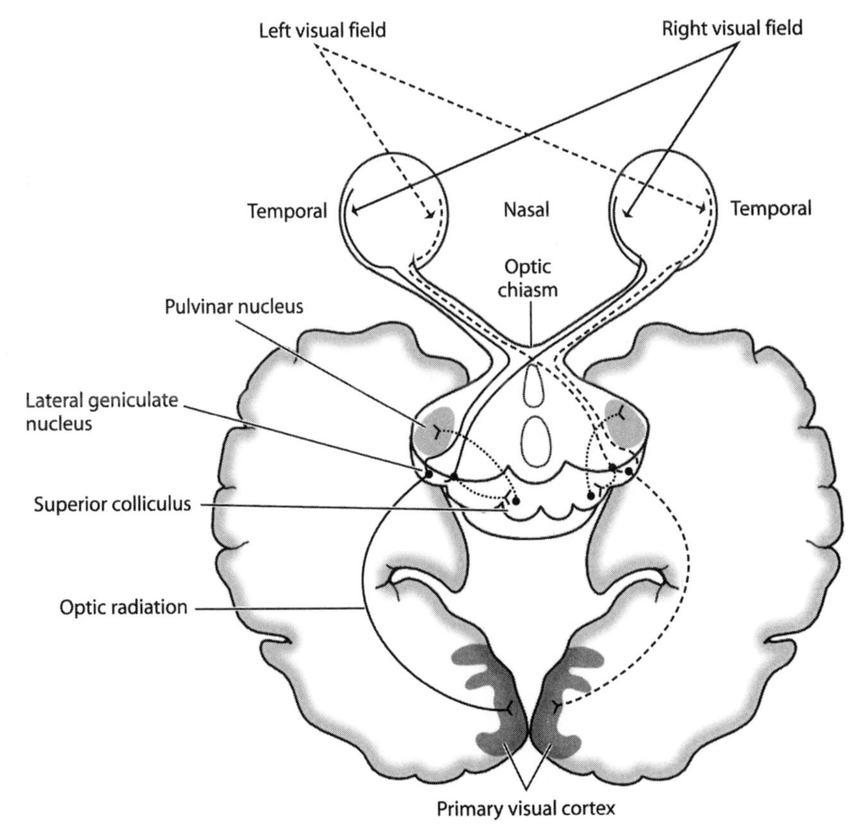

Figure 6

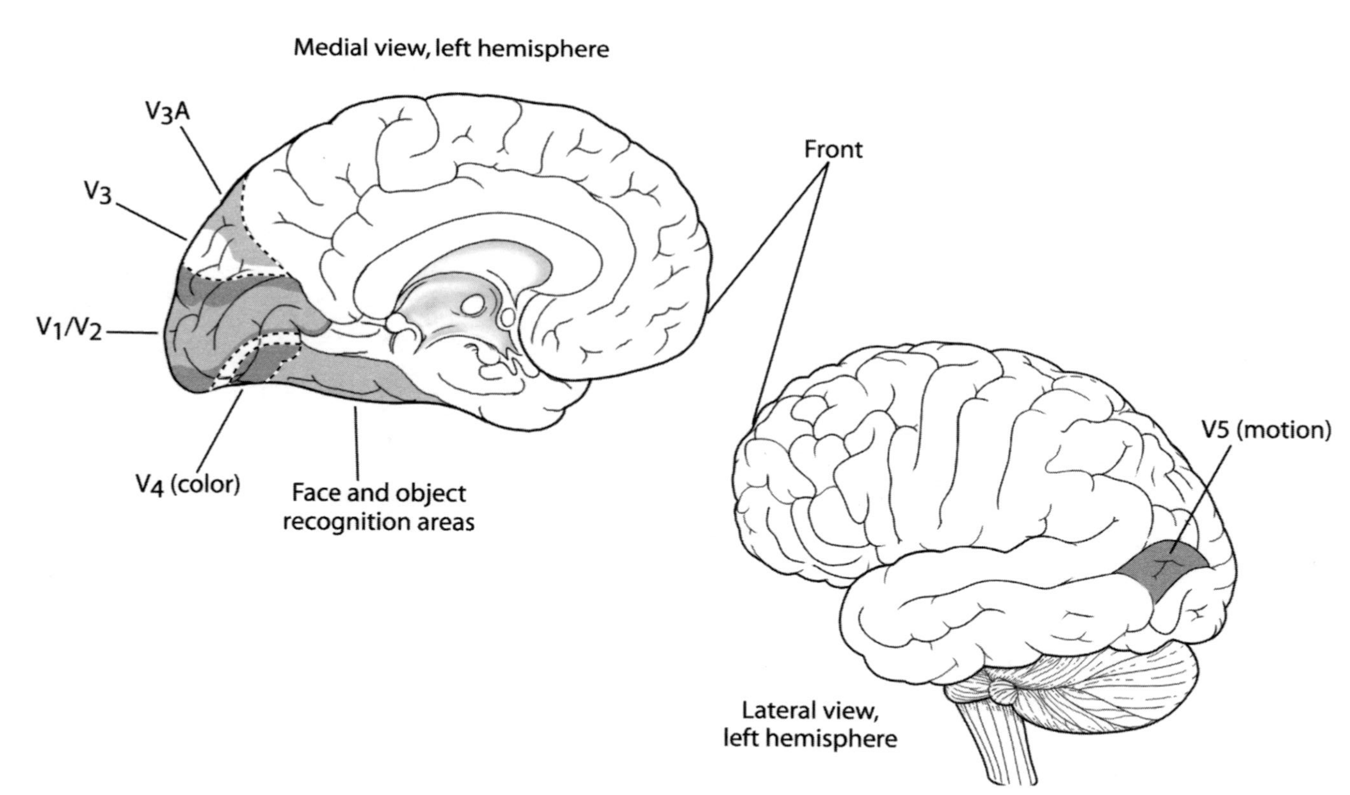
Medial view, left hemisphere
V3A
V3
V1/V2
V4 (color)
Face and object recognition areas
Front
V5 (motion)
Lateral view, left hemisphere

Figure 7

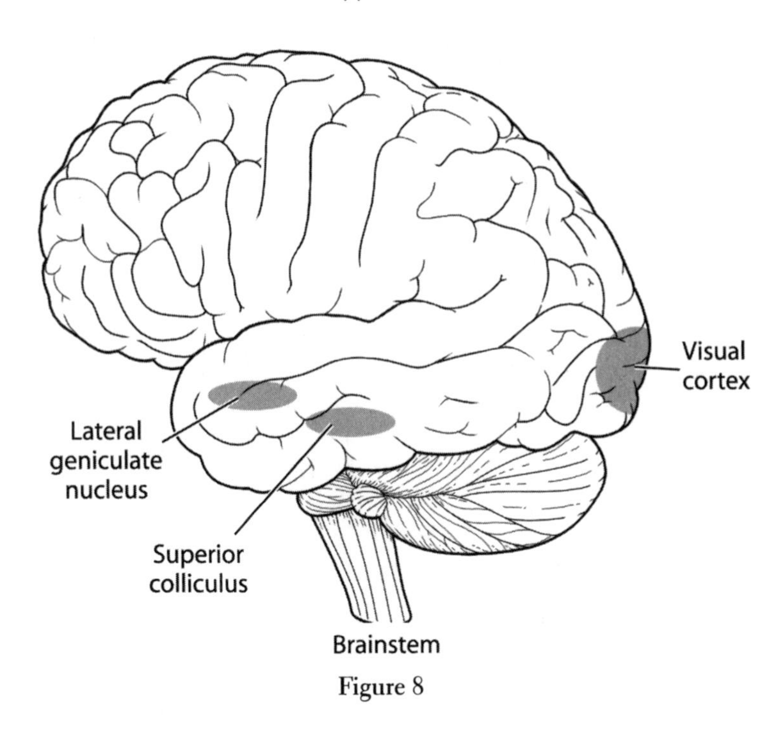

Figure 8

on the functions of the V-designated regions, as well as the lateral geniculate nucleus and superior colliculus) (Figures 7 and 8).

The temporal lobe performs several functions, including hearing, perception, and some kinds of memory. An area of the temporal cortex about as big as a poker chip is responsible for hearing. It's called the auditory cortex. Damage to the left temporal lobe can cause loss of language (Figure 9).

NAMES OF BRAIN STRUCTURES

One of the problems with learning about the brain is that different researchers use different names for the same brain structures. For example, the nucleus accumbens may also be called the ventral striatum. It's a part of a larger structure called the striatum, which is also called the caudate putamen (if rats are being studied). In humans, the caudate can be differentiated from the putamen, and together they are sometimes called the basal ganglia, particularly by people who study movement. The trouble is that the basal ganglia include other structures besides the striatum, such as substantia nigra, which is where the dopamine-producing neurons die in people who have Parkinson's disease. In Chapter 12,

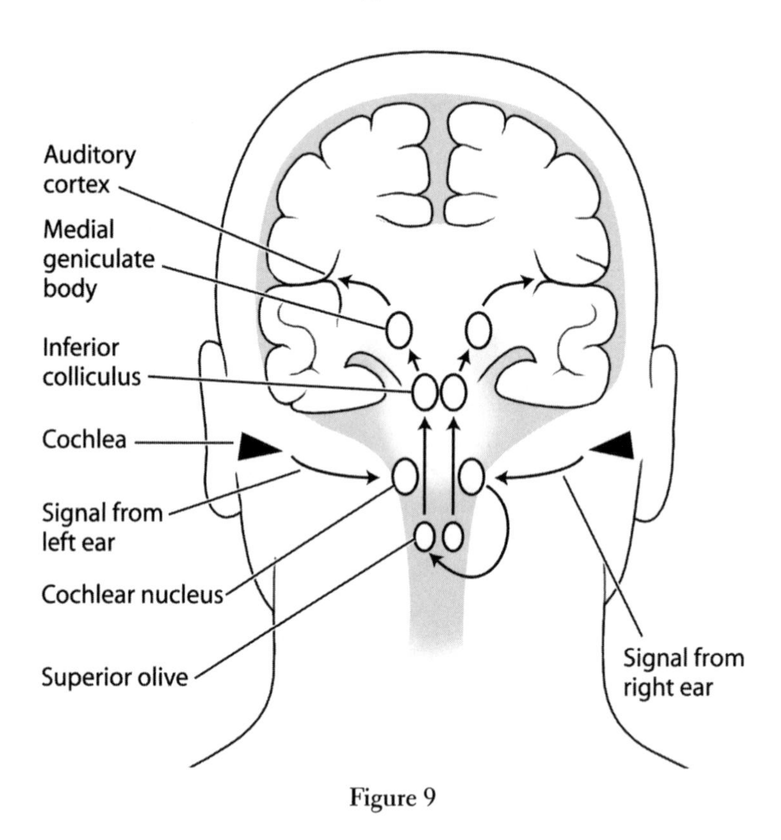

Figure 9

you read about work on the reward centers of the ventral pallidum. It's also called the globus pallidus and is yet another part of the basal ganglia, as shown in Figure 10.

Confused? Don't feel alone. Even the experts can't agree, and they sometimes have trouble sharing their results because they are using the same terms to mean different things or different terms to mean the same thing. That's up to the researchers to sort out, and they will—probably over the next decade or two. In the meantime, the easiest thing for the rest of us to do is use our favorite search engine to locate diagrams of brain structure, relying on Web sources that come from governmental and educational institutions (the dot.coms don't always get it right), carefully comparing what we find in several sites (even the most academic sources don't always show the same thing). In the end, it's important to remember that brain mapping is still an inexact science, and that the regions mentioned in this book and elsewhere give us only a general idea of the brain's complex architecture.

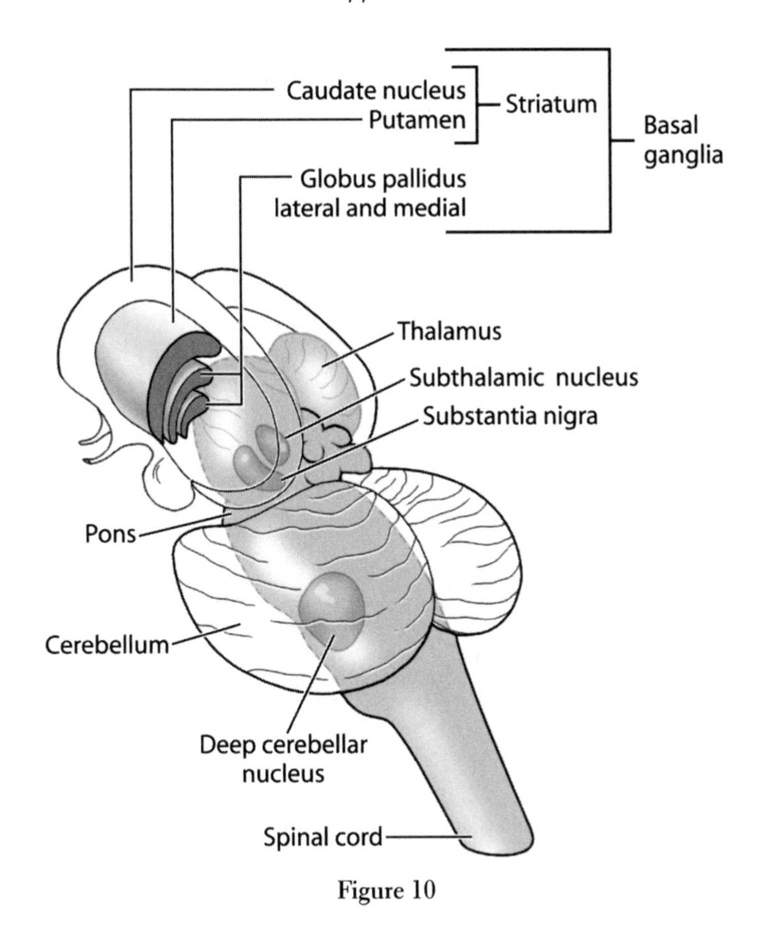

Figure 10

THE NERVOUS SYSTEM

The nervous system has two parts: the central nervous system, which consists of the brain and spinal cord; and the peripheral nervous system, which carries messages to and from the central nervous system via twelve pairs of cranial nerves and thirty-one pairs of spinal nerves (Figure 11). The twelve sets of nerve fibers that connect directly to the brain (not through the spinal cord) are often referred to by Roman numerals I to XII or by their names (in order): olfactory, optic, oculomotor, trochlear, trigeminal, abducens, facial, vestibulocochlear, glossopharyngeal, vagus, accessory, and hypoglossal nerves.

Cells in the peripheral nervous system include sensory neurons that respond to changes in the environment (touch, heat, pain, light, sound vibrations, and more) and send messages to the brain. Motor neurons are also part of the peripheral nervous system. They cause a muscle to contract or a gland to secrete in

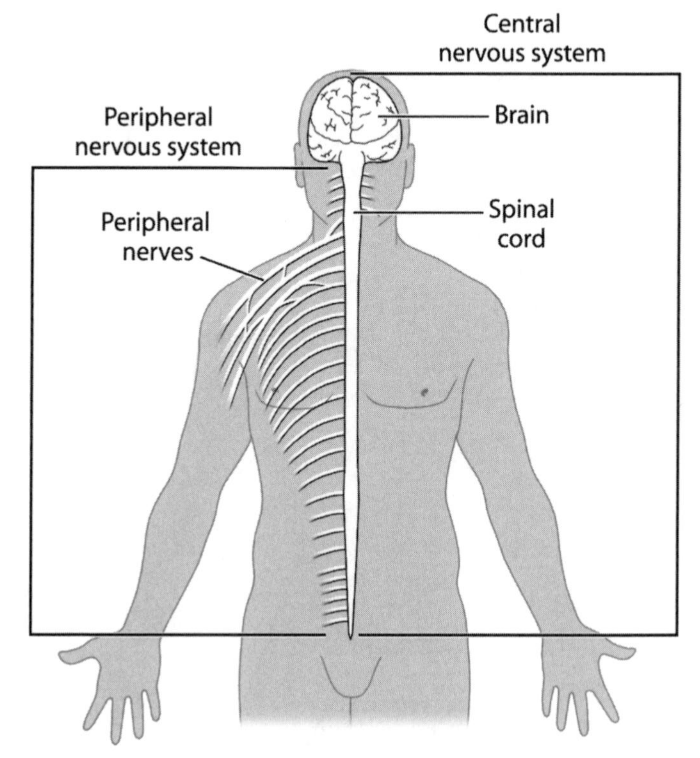

Figure 11

response to a signal from the central nervous system. Interneurons, a third type of nerve cell, are the "go-betweens." In the spinal cord, one of the jobs of interneurons is to pass messages between sensory and motor neurons (Figure 11).

Some peripheral nerves carry sensory information from the body to the brain via the spinal cord. Others carry signals from the brain to the body, again using the spinal cord as an "information highway." Your brain knows when your back itches because electrical impulses travel along nerve cells to the brain. There the brain interprets the input as an itch and sends the signal to scratch.

The part of the peripheral nervous system that controls the internal organs is called the autonomic nervous system. It regulates the heart muscle and the smooth muscle in internal organs such as the intestine, bladder, and uterus. It has two divisions. The sympathetic division usually promotes rapid change in body systems, such as making the heart beat faster when danger looms. The parasympathetic nervous system has the opposite effect; it tends to slow things down. The two systems work in opposition and allow the body to maintain a healthy balance.

NEURONS

In most ways, neurons are the same as other cells in the body. They have a nucleus, a membrane around the outside, and mitochondria where energy is released to power the cell. (That's where oxygen is used to "burn" food and get the energy out.) But neurons have some unique structures. The axon is a single fiber at one end that carries messages away from the cell body. At the other end of the neuron are short fibers called dendrites. They receive information from the axons of other neurons.

Neurons do not touch, so when scientists speak of connections, they don't mean actual physical hook-ups. They mean pathways for sending signals. The transmission of impulses between neurons that do not touch is achieved by chemical messengers.

The axon of one neuron, a dendrite of another, and the gap between them are collectively called a synapse. Axons release chemicals called neurotransmitters into the synaptic gap. Neurotransmitters may be amino acids, proteins, or even gases (Figure 12).

Neurotransmitter molecules move across the gap the same way a drop of food coloring diffuses in water. When a neurotransmitter molecule reaches the dendrite of another neuron, it attaches to a site called a receptor. Receptors are large protein molecules that stick out from the membrane. They have a particular shape that fits only the neurotransmitter they receive. Think of a lock and key. Only one key fits one particular lock. When enough neurotransmitter molecules latch onto their receptors, a nervous impulse begins (or stops) in the receiving neuron.

Something like ten *quadrillion* synapses (that's a thousand trillion or a 1 with fifteen zeroes after it) connect the neurons in the brain.[12] It might seem that such vast numbers would make the brain an incomprehensible tangle of competing electrical signals, but not so. The brain is remarkably efficient in using relatively few signals to direct the great repertoire of senses, thoughts, and actions that humans and other animals can achieve. Recent research suggests, for example, that a mouse can tell how fast its body is moving from the firing of a mere one hundred of its motion-sensitive synapses.[13]

THE NEURAL IMPULSE

Nerve impulses are sometimes compared to electrical wires, with the messages traveling along them like a current. That comparison isn't exactly right, but it is true that electrical potentials change across the outer membranes of neurons. Here's what happens: A neuron at rest has a slightly negative charge inside it, compared with a positive charge outside. When an excitatory neurotransmitter attaches to a receptor, it causes tiny channels in the membrane of the axon to open. The channels let charged particles flood in, thus changing the charge along

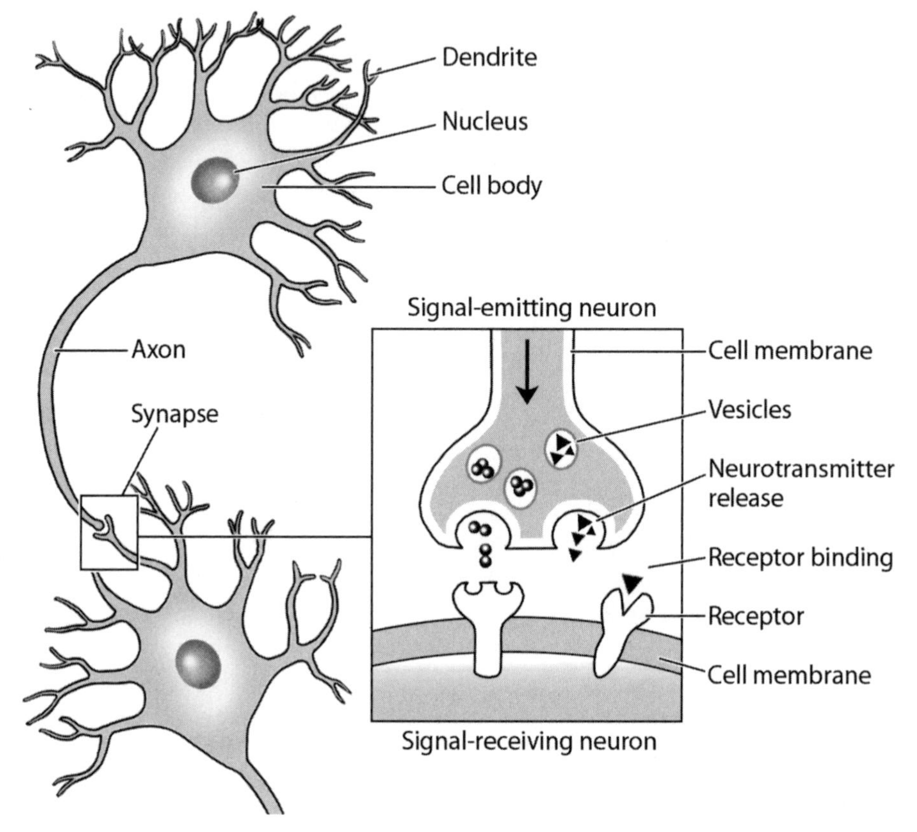

Figure 12

the membrane. For a tiny fraction of a second, the inside of the cell becomes positive and the outside becomes negative. This reversal in one tiny area creates a current that affects the membrane farther along. Channels there open, and charged particles flow into another section of the axon, then another, then another. That's how the impulse travels the length of the axon. Inhibitory neurotransmitters make the inside of the axon even more negatively charged than usual, so it is harder to excite.

A nerve cell continues firing as long as enough excitatory neurotransmitter molecules stay bound to it. (In the same way, receiving neurons won't fire when large numbers of inhibitory neurotransmitter molecules are bound to them.) Also, habituation can occur. The longer the excitatory stimulus continues, the less often neurons react.

In time, neurotransmitters either break down or are reabsorbed. Why don't the channels just stay open and neurons keep firing forever? Because a mechanism in the cell membrane called an ion pump pushes charged particles of the axon in the same way that a water pump clears a flooded basement.

Notes

PREFACE: FALLING IN LOVE WITH SCIENCE

1. Richard M. Restak, *The Brain.* New York: Bantam, 1984, 349.

CHAPTER 1: LIFE WITHOUT TOUCH

1. All quotes from Chuck Aoki and Jennifer Nelson are personal communications.
2. Felicia B. Axelrod and Gabrielle Gold-von Simson, "Hereditary Sensory and Autonomic Neuropathies: Types II, III, and IV," *Orphanet Journal of Rare Diseases* (2007) 2: 39. http://www.pubmedcentral.nih.gov/articlerender.fcgi?artid=2098750. Accessed June 3, 2008.
3. Larry Gray, Lisa Watt, and Elliot M. Blass, "Skin-to-Skin Contact Is Analgesic in Healthy Newborns," *Pediatrics* (January 2000) 105(1): e14.

CHAPTER 2: IN FROM THE COLD

1. Andrea M. Peier, Aziz Moqrich, Anne C. Hergarden, Alison J. Reeve, David A. Andersson, Gina M. Story, Taryn J. Earley, Ilaria Dragoni, Peter McIntyre, Stuart Bevan, and Ardem Patapoutian, "A TRP Channel that Senses Cold Stimuli and Menthol." *Cell* (March 8, 2002) 108(5): 705–15.
2. G. Reid and M. L. Flonta, "Ion Channels Activated by Cold and Menthol in Cultured Rat Dorsal Root Ganglion Neurones," *Neuroscience Letter* (May 17, 2002) 324(2): 164–8.
3. Carline Seydel, "How Neurons Know That It's C-c-c-c-cold Outside," *Science* (February 22, 2002) 295(5559): 1451.

4. David D. McKemy, Werner M. Neuhausser, and David Julius, "Identification of a Cold Receptor Reveals a General Role for TRP Channels in Thermosensation," *Nature* (March 7, 2002) 416: 52–8.

5. Peier et al., 2002.

6. Seydel, 2002.

7. R. F. Johnstone, W. K. Dolen, and H. M. Hoffman, "A Large Kindred with Familial Cold Autoinflammatory Syndrome," *Annals of Allergy Asthma and Immunology* (February 2003) 90(2): 233–7.

8. Doherty quoted in "Dermatology Team Finds Treatment for Rare 'Life-ruining' Condition," Press release from Dalhousie University (March 11, 2008).

9. J. B. Ross, L. A. Finlayson, P. J. Klotz, R. G. Langley, R. Gaudet, K. Thompson, S. M. Churchman, M. F. McDermott, and P. N. Hawkins, "Use of Anakinra (Kineret) in the Treatment of Familial Cold Autoinflammatory Syndrome with a 16-month Follow-up," *Journal of Cutaneous Medicine and Surgery* (January–February 2008) 12(1): 8–16.

10. Félix Viana, Elvira de la Peña, and Carlos Belmonte, "Specificity of Cold Thermotransduction Is Determined by Differential Ionic Channel Expression," *Nature Neuroscience* (2002) 5: 254–60.

11. Wei Cheng, Fan Yang, Christina L. Takanishi, and Jie Zheng, "Thermosensitive TRPV Channel Subunits Coassemble into Heteromeric Channels with Intermediate Conductance and Gating Properties," *Journal of General Physiology* (March 2007) 129(3): 191–207.

CHAPTER 3: ON WHICH SIDE IS YOUR BREAD BUTTERED?

1. E. Roy and C. MacKenzie, "Handedness Effects in Kinesthetic Spatial Location Judgements," *Cortex* (1978) 14: 250–8.

2. Daniel J. Goble and Susan H. Brown, "Task-Dependent Asymmetries in the Utilization of Proprioceptive Feedback for Goal-Directed Movement," *Experimental Brain Research* (2007)180: 693–704.

3. Daniel J. Goble and Susan H. Brown, "Upper Limb Asymmetries in the Matching of Proprioceptive versus Visual Targets," *Journal of Neurophysiology* (June 2008) 99(6):3063–74.

4. S. G. Kim, J. Ashe, K. Hendrich, J. M. Ellermann, H. Merkle, K. Uğurbil, and A. P. Georgopoulos, "Functional Magnetic Resonance Imaging of Motor Cortex: Hemispheric Asymmetry and Handedness," *Science* (1993) 261(5121): 615–7.

5. G. Rains and B. Milner, "Right-Hippocampal Contralateral-Hand Effect in the Recall of Spatial Location in the Tactual Modality." *Neuropsychologia* (1994) 32: 1233–42.

6. Eiichi Naito1, Per E. Roland, Christian Grefkes, H. J. Choi, Simon Eickhoff, Stefan Geyer, Karl Zilles, and H. Henrik Ehrsson, "Dominance of the Right Hemisphere and Role of Area 2 in Human Kinesthesia," *Journal of Neurophysiology* (2005) 93: 1020–34.

7. A. Butler, G. Fink, C. Dohle, G. Wunderlich, L. Tellmann, R. Seitz, K. Zilles, and H. Freund, "Neural Mechanisms Underlying Reaching for Remembered Targets Cued Kinesthetically or Visually in Left or Right Hemispace," *Human Brain Mapping* (2004) 21: 165–77.

CHAPTER 4: PAIN AND THE PLACEBO EFFECT

1. Andrea Dalton and Meredyth Daneman, "Social Susceptibility to Central and Peripheral Misinformation," *Memory* (May 2006) 14(4): 486–501.

2. Estimate from the Society for Neuroscience.

3. "Cannabinoids and Pain," *Brain Briefings*, Society for Neuroscience, March 1999.

4. Jon-Kar Zubieta, Mary M. Heitzeg, Yolanda R. Smith, Joshua A. Bueller, Ke Xu, Yanjun Xu, Robert A. Koeppe, Christian S. Stohler, and David Goldman, "COMT *val*^{158}*met* Genotype Affects μ-Opioid Neurotransmitter Responses to a Pain Stressor," *Science* (February 21, 2003) 299(5610): 1240–3.

5. S.D. Dorn, T. J. Kaptchuk, J. B. Park, L. T. Nguyen, K. Canenguez, B. H. Nam, K. B. Woods, L.A. Conboy, W. B. Stason, and A. J. Lembo, "A Meta-analysis of the Placebo Response in Complementary and Alternative Medicine Trials of Irritable Bowel Syndrome," *Neurogastroenterology and Motility* (August 2007) 19(8): 630–7.

6. M. E. Kemeny, L. J. Rosenwasser, R. A. Panettieri, R. M. Rose, S. M. Berg-Smith, and J. N. Kline, "Placebo Response in Asthma: A Robust and Objective Phenomenon," *Journal of Allergy and Clinical Immunolog y* (June 2007) 119(6): 1375–81.

7. For a good overview of these and other findings, see Steve Stewart-Williams, "The Placebo Puzzle: Putting Together the Pieces," *Health Psychology* (March 2004) 23(2): 198–206.

8. Fabrizio Benedetti, Claudia Arduino, and Martina Amanzio, "Somatotopic Activation of Opioid Systems by Target-Directed Expectations of Analgesia," *Journal of Neuroscience* (May 1, 1999) 19(9): 3639–48.

9. J. D. Levine, N. C. Gordon, and H. L. Fields, "The Mechanisms of Placebo Analgesia," *Lancet* (1978) 2: 654–7.

10. J. J. Lipman, B. E. Miller, K. S. Mays, M. N. Miller, W. C. North, and W. L. Byrne, "Peak B Endorphin Concentration in Cerebrospinal Fluid: Reduced in Chronic Pain Patients and Increased During the Placebo Response," *Psychopharmacology* (September 1990) 102(1): 112–6.

11. M. Amanzio and F. Benedetti, "Neuropharmacological Dissection of Placebo Analgesia: Expectation-Activated Opioid Systems versus Conditioning-Activated Specific Subsystems," *Journal of Neuroscience* (January 1, 1999) 19(1): 484–94.

12. Fabrizio Benedetti, Antonella Pollo, Leonardo Lopiano, Michele Lanotte, Sergio Vighetti, and Innocenzo Rainero, "Conscious Expectation and Unconscious Conditioning in Analgesic, Motor, and Hormonal Placebo/Nocebo Responses," *Journal of Neuroscience* (May 15, 2003) 23(10): 4315–23.

13. P. Petrovic, T. Dietrich, P. Fransson, J. Andersson, K. Carlsson, and M. Ingvar, "Placebo in Emotional Processing-Induced Expectations of Anxiety Relief Activate a Generalized Modulatory Network," *Neuron* (June 16, 2005) 46(6): 957–69.

14. M. N. Baliki, P. Y. Geha, A. V. Apkarian, and D. R. Chialvo, "Beyond Feeling: Chronic Pain Hurts the Brain, Disrupting the Default-Mode Network Dynamics," *Journal of Neuroscience* (February 6, 2008) 28(6): 1398–403.

15. All quotes from Dante Chialvo are personal communications.

16. Raúl de la Fuente-Fernández, Thomas J. Ruth, Vesna Sossi, Michael Schulzer, Donald B. Calne, and A. Jon Stoessl, "Expectation and Dopamine Release: Mechanism of the Placebo Effect in Parkinson's Disease," *Science* (August 10, 2001) 293(5532): 1164–6.

17. Thomas Luparello, Harold A. Lyons, Eugene R. Bleecker, and E. R. McFadden, "Influences of Suggestion on Airway Reactivity in Asthmatic Subjects," Psychosomatic *Medicine* (1968) 30: 819–25.

18. J-K. Zubieta, J. A. Bueller, L. R. Jackson, D. J. Scott, Y. Xu, R. A. Koeppe, T. E. Nichols, and C. S. Stohler, "Placebo Effects Mediated by Endogenous Opioid Activity on Mu-opioid Receptors," *Journal of Neuroscience* (2005) 25(34): 7754–62.

CHAPTER 5: NEMATODES, HAPTICS, AND BRAIN–MACHINE INTERFACES

1. Robert O'Hagan and Martin Chalfie, "Mechanosensation in *Caenorhabditis elegans*," *International Review of Neurobiology* (2006) 69: 169–203.

2. Laura Bianchi, "Mechanotransduction: Touch and Feel at the Molecular Level as Modeled in *Caenorhabditis elegans*," *Molecular Neurobiology* (December 2007) 36(3): 254–71.

3. M. Chalfie and J. Sulston, "Developmental Genetics of the Mechanosensory Neurons of *Caenorhabditis elegans*," *Developmental Biology* (1981) 82: 358–70.

4. Bianchi, 2007.

5. Rajesh Babu Sekar and Ammasi Periasamy, "Fluorescence Resonance Energy Transfer (FRET) Microscopy Imaging of Live Cell Protein Localizations," *Journal of Cell Biology* (March 3, 2003) 160(5): 629–33.

6. Bianchi, 2007.

7. Miriam B. Goodman and Erich M. Schwarz, "Transducing Touch in *Caenorhabditis elegans*," *Annual Review of Physiology* (2003) 65: 429–52.

8. K. Dandekar, B. I. Raju, and M. A. Srinivasan, "3-D Finite-Element Models of Human and Monkey Fingertips to Investigate the Mechanics of Tactile Sense," *Journal of Biomechanical Engineering* (October 2003) 125: 682–91.

9. "MIT and London Team Report First Transatlantic Touch," MIT News, October 28, 2002. http://web.mit.edu/newsoffice/nr/2002/touchlab3.html. Accessed July 14, 2008.

10. Mandayam A Srinivasan, "What Is Haptics?" http://www.sensable.com/documents/documents/what_is_haptics.pdf. Accessed July 14, 2008.

11. Gil Yosipovitch, Yozo Ishiuji, Tejesh S. Patel, Maria Isabel Hicks, Yoshitetsu Oshiro, Robert A. Kraft, Erica Winnicki, and Robert C. Coghill, "The Brain Processing of Scratching," *Journal of Investigative Dermatology* (July 2008) 128(7): 1806–11.

12. "Research Suggests Why Scratching Is So Relieving," Press release from Wake Forest University Baptist Medical Center, January 31, 2008, available at http://www.eurekalert.org/pub_releases/2008–01/wfub-rsw012808.php. Accessed March 3, 2008.

13. Srinivasan, 2008.

14. M. Velliste, S. Perel, M. C. Spalding, A. S. Whitford, and A. B. Schwartz, "Cortical Control of a Prosthetic Arm for Self- Feeding," *Nature* (June 19 2008) 453: 1098–1109.

CHAPTER 6: IT'S VALENTINE'S DAY . . . SNIFF!

1. Michelle L. Schlief and Rachel I. Wilson, "Olfactory Processing and Behavior Downstream from Highly Selective Receptor Neurons," *Nature Neuroscience* (May 1, 2007) 10: 623–30.

2. Rachel I. Wilson, "Neural Circuits Underlying Chemical Perception," *Science* (October 26, 2007) 318: 584–5.

3. Eric H. Holbrook and Donald A. Leopold, "Anosmia: Diagnosis and Management," *Current Opinion in Otolaryngology & Head and Neck Surgery* (2003) 11: 54–60.

4. Z. Zou and L. B. Buck, "Combinatorial Effects of Odorant Mixes in Olfactory Cortex," *Science* (March 10, 2006) 311(5766): 1477–81.

5. Z. Zou, F. Li, and L. B. Buck, "Odor Maps in the Olfactory Cortex," *Proceedings of the National Academy of Sciences* (May 24, 2005) 102(21): 7724–9.

6. Leslie B. Vosshall, "How the Brain Sees Smells," *Developmental Cell* (November 2001) 1(5): 588–90.

7. Friedrich quoted in "When the Nose Doesn't Know: Cornell Scientists Develop Standard Spectrum of Smell to Separate 'Ahhh' from 'Ugghh,'" Press release from Cornell University, March 26, 2000. http://www.news.cornell.edu/releases/March00/ACS.SmellStandards.bpf.html. Accessed March 22, 2008.

8. M. Tsunozaki, S. H. Chalasani, and C. I. Bargmann, "A Behavioral Switch: cGMP and PKC Signaling in Olfactory Neurons Reverses Odor Preference in C. *elegans*," *Neuron* (September 25, 2008) 59(6): 959–71.

9. Weihong Lin, Tatsuya Ogura, Robert F. Margolskee, Thomas E. Finger, and Diego Restrepo, "TRPM5-expressing Solitary Chemosensory Cells Respond to Odorous Irritants," *Journal of Neurophysiology* (March 1, 2008) 99(3): 1451–60.

10. D. Rinberg, A. Koulakov, and A. Gelperin, "Speed Accuracy Tradeoff in Olfaction," *Neuron* (August 3, 2006) 51 (3): 351–8.

11. Wen Li, Erin Luxenberg, Todd Parrish, and Jay A. Gottfried, "Learning to Smell the Roses: Experience-Dependent Neural Plasticity in Human Piriform and Orbitofrontal Cortices," *Neuron* (December 21, 2006) 52(6): 1097–1108.

12. Gottfried quoted in "Researchers Discover How We Differentiate Smells," *Medical News Today*, December 27, 2006, http://www.medicalnewstoday.com/articles/59697.php. Accessed September 7, 2008.

13. P. N. Steinmetz, A. Roy, P. J. Fitzgerald, S. S. Hsiao, K. O. Johnson, and E. Niebur, "Attention Modulates Synchronized Neuronal Firing in Primate Somatosensory Cortex," *Nature* (March 9, 2000) 404: 187–90.

14. N. Sobel, V. Prabhakaran, J. E. Desmond, G. H. Glover, R. L. Goode, E. V. Sullivan, and J. D. E. Gabrieli, "Sniffing and Smelling: Separate Subsystems in the Human Olfactory Cortex," *Nature* (March 19, 1998) 392: 282–6.

15. Steven D. Munger, Andrew P. Lane, Haining Zhong, Trese Leinders-Zufall, King-Wai Yau, Frank Zufall, and Randall R. Reed, "Central Role of the CNGA4 Channel Subunit in Ca2+-Calmodulin-Dependent Odor Adaptation," *Science* (December 7, 2001) 294: 2172–5.

16. Yuan Gao1 and Ben W Strowbridge, "Long-Term Plasticity of Excitatory Inputs to Granule Cells in the Rat Olfactory Bulb," *Nature Neuroscience* Published online: May 3, 2009, at http://www.nature.com/neuro/journal/vaop/ncurrent/abs/nn.2319.html.

17. Jess Porter, Brent Craven, Rehan M Khan, Shao-Ju Chang, Irene Kang, Benjamin Judkewitz, Jason Volpe, Gary Settles, and Noam Sobel, "Mechanisms of Scent-Tracking in Humans," *Nature Neuroscience* (December 17, 2007) 10: 27–9.

CHAPTER 7: DO ODORS DANCE IN YOUR NOSE?

1. "Luca Turin on Etat Libre d'Orange in NZZ Folio {Fragrant Reading}," http://www.mimifroufrou.com/scentedsalamander/2007/08/luca_turin_on_etat_libre_doran.html. Accessed September 20, 2008

2. Chandler Burr, *The Emperor of Scent: A True Story of Perfume and Obsession*. New York: Random House, 2003: 10.

3. Avery Gilbert, "The Emperor's New Theory," (A Review of *The Emperor of Scent* by Chandler Burr), *Nature Neuroscience* (April 2003) 6(4): 1.

4. Luca Turin, *The Secret of Scent*. New York: HarperCollins, 2006: 133.

5. Luca Turin, "A Spectroscopic Mechanism for Primary Olfactory Reception," *Chemical Senses* (December 1996) 21(6): 773–91.

6. Stephen Hill, "Sniff 'n' Shake," *NewScientist* (January 3, 1998) 157(2115): 34–7.

7. Ibid.

8. L. J. W. Haffenden, V. A. Yaylayan, and J. Fortin, "Investigation of Vibrational Theory of Olfaction with Variously Labelled Benzaldehydes," *Food Chemistry* (2001) 73: 67–72.

9. Gilbert, 2003.

10. Andreas Keller and Leslie B. Vosshall, "A Psychophysical Test of the Vibration Theory of Olfaction," *Nature Neuroscience* (April 2004) 7(4): 337–8.

11. "Testing a Radical Theory," *Nature Neuroscience* (April 2004) 7(4): 315.

12. Turin, *Secret of Scent*, 188.

13. Shin-ya Takane and John B. O. Mitchell, "A Structure–Odour Relationship Study Using EVA Descriptors and Hierarchical Clustering," *Organic and Biomolecular Chemistry* (November 21, 2004) 2(22): 3250–5.

14. Philip Ball, "Rogue Theory of Smell Gets a Boost," published online December 7, 2006, http://www.nature.com/news/2006/061204/full/news061204-10.html. Accessed March 2, 2008.

15. Jennifer C. Brookes, Filio Hartoutsiou, A. P. Horsfield, and A. M. Stonehamx, "Could Humans Recognize Odor by Phonon Assisted Tunneling?" *Physical Review Letters* (January 19, 2007) 98(3): 38101–4.

16. Zeeya Merali, "Nose Model Gets a Cautious Sniff," *NewScientist* (December 9, 2006): 12.

17. London Centre for Nanotechnology, "Rogue Odour Theory Could Be Right." *ScienceDaily*. (February 5, 2007), http://www.sciencedaily.com—/releases/2007/02/070204162541.htm. Accessed March 2, 2008.

CHAPTER 8: LIFE WITHOUT SCENT

1. All quotes from Theresa Dunlop are personal communications.

2. Matthieu Louis, Thomas Huber, Richard Benton, Thomas P. Sakmar, and Leslie B. Vosshall, "Bilateral Olfactory Sensory Input Enhances Chemotaxis Behavior," *Nature Neuroscience* (February 1, 2008) 11: 187–99.

3. A. Brämerson, L. Johansson, L. Ek, S. Nordin, and M. Bende, "Prevalence of Olfactory Dysfunction: The Skövde Population-Based Study," *Laryngoscope* (April 2004) 114(4): 733–7.

4. C. Murphy, C. R. Schubert, K. J. Cruickshanks, B. E. Klein, R. Klein, and D. M. Nondahl, "Prevalence of Olfactory Impairment in Older Adults," *Journal of the American Medical Association* (November 13, 2002) 288 (18): 2307–12.

5. Andreas F. P. Temmel, Christian Quint, Bettina Schickinger-Fischer, Ludger Klimek, Elisabeth Stoller, and Thomas Hummel, "Characteristics of Olfactory Disorders in Relation to Major Causes of Olfactory Loss," *Archives of Otolaryngology and Head and Neck Surgery* (June 2002) 128: 635–41.

6. I. J. Swann, B. Bauza-Rodrigue, R. Currans, J. Riley, and V. Shukla, "The Significance of Post-Traumatic Amnesia as a Risk factor in the Development of Olfactory Dysfunction Following Head Injury," *Emergency Medicine Journal* (2006) 23: 618–21.

7. Brämerson et al., 2004.

8. Eric H. Holbrook and Donald A. Leopold, "Anosmia: Diagnosis and Management," *Current Opinion in Otolaryngology & Head and Neck Surgery* (2003) 11: 54–60.

9. Antje Haehner, Antje Rodewald, Johannes C. Gerber, and Thomas Hummel, "Correlation of Olfactory Function with Changes in the Volume of the Human Olfactory Bulb," *Archives of Otolaryngology Head and Neck Surgery* (2008) 134(6): 621–4.

10. X. Grosmaitre, L. C. Santarelli, J. Tan, M. Luo, and M. Ma, "Dual Functions of Mammalian Olfactory Sensory Neurons as Odor Detectors and Mechanical Sensors," *Nature Neuroscience* (March 2007) 10(3): 348–54.

11. Ma quoted in "Explaining Why We Smell Better When We Sniff," Press release from the University of Pennsylvania School of Medicine, March 13, 2007, http://www.upenn.edu/researchatpenn/article.php?1206&sci. Accessed September 8, 2008.

12. C. T. Leung, P. A. Coulombe, and R. R. Reed, "Contribution of Olfactory Neural Stem Cells to Tissue Maintenance and Regeneration," *Nature Neuroscience* (June 2007) 10(6): 673–4.

CHAPTER 9: THE SWEATY SCENT OF SEX

1. Stephen D. Liberles and Linda S. Buck, "A Second Class of Chemosensory Receptors in the Olfactory Epithelium," *Nature* (August 10, 2006) 442(10): 645–50.

2. According to http://honeymoons.about.com.

3. Liberles and Buck, 2006, 648.

4. Leslie B. Vosshall, "Olfaction: Attracting Both Sperm and the Nose," *Current Biology* (November 9, 2004) (14): R918–20.

5. Suma Jacob, Martha K. McClintock, Bethanne Zelano, and Carole Ober, "Paternally Inherited HLA Alleles Are Associated with Women's Choice of Male Odor," *Nature Genetics* (February 2002) 50: 175–9.

6. Rockefeller University, "Male Body Odor Can Stink Like Urine or Have A Pleasant Vanilla Smell, Depending On One Gene." *ScienceDaily*. (September 17, 2007), http://www.sciencedaily.com /releases/2007/09/070916143523.htm. Accessed March 8, 2008.

7. Public Library of Science, "Why Are Some People Oblivious to the 'Sweaty' Smell of a Locker Room?" *ScienceDaily* (November 1, 2007), http://www.sciencedaily.com / releases/2007/10/071030080645.htm. Accessed March 8, 2008

8. P. S. Santos, J. A. Schinemann, J. Gabardo, and Mda G. Bicalho, "New Evidence That the MHC Influences Odor Perception in Humans: A Study with 58 Southern Brazilian Students," *Hormones and Behavior* (April 2005) 47(4): 384–8.

9. B. I. Grosser, L. Monti-Bloch, C. Jennings-White, and D. L. Berliner, "Behavioral and Electrophysiological Effects of Androstadienone, a Human Pheromone," *Psychoneuroendocrinology* (April 2000) 25(3): 289–99.

10. Ivanka Savic, Hans Berglund, and Per Lindström, "Brain Response to Putative Pheromones in Homosexual Men," *Proceedings of the National Academy of Sciences* (May 17, 2005) 102(20): 7356–61.

11. S Jacob and M. K McClintock, "Psychological State and Mood Effects of Steroidal Chemosignals in Women and Men," *Hormones and Behavior* (2000) 37(1): 57–78.

12. C. Wyart, W. W. Webster, J. H. Chen, S. R. Wilson, A. McClary, R. M. Khan, and N. Sobel, "Smelling a Single Component of Male Sweat Alters Levels of Cortisol in Women," *Journal of Neuroscience* (February 2007) 27(6): 1261–5.

13. Ibid.

14. J. N. Lundström and M. J.Olsson, "Subthreshold Amounts of Social Odorant Affect Mood, But Not Behavior, in Heterosexual Women When Tested by a Male, But Not a Female, Experimenter," *Biological Psychology* (December 2005) 70(3): 197–204.

15. M. Bensafi, W. M. Brown, R. Khan, B. Levenson, and N. Sobel, "Sniffing Human Sex-Steroid Derived Compounds Modulates Mood, Memory and Autonomic Nervous System Function in Specific Behavioral Contexts," *Behavioral Brain Research* (June 4, 2004) 152(1): 11–22.

16. S. Jacob, L. H. Kinnunen, J. Metz, M. Cooper, and M. K. McClintock, "Sustained Human Chemosignal Unconsciously Alters Brain Function," *NeuroReport* (2001) 12(11): 2391–4.

17. J. N. Lundström, M. J. Olsson, B. Schaal, and T. Hummel, "A Putative Social Chemosignal Elicits Faster Cortical Responses Than Perceptually Similar Odorants," *Neuroimage* (May 1, 2006) 30(4): 1340–6.

18. J. N. Lundström, J. A. Boyle, R. J. Zatorre, and M. Jones-Gotman, "Functional Neuronal Processing of Body Odors Differs from that of Similar Common Odors," *Cerebral Cortex* (June 2008) 18(6): 1466–74.

19. I. Savic, H. Berglund, B. Gulyas, and P. Roland, "Smelling of Odorous Sex Hormone-Like Compounds Causes Sex-Differentiated Hypothalamic Activations in Humans," *Neuron* (2001) 31(4): 661–8.

20. Jakob C. Mueller, Silke Steiger, Andrew E. Fidler, and Bart Kempenaers, "Biogenic Trace Amine–Associated Receptors (TAARs) Are Encoded in Avian Genomes: Evidence and Possible Implications," *Journal of Heredity* (March–April 2008) 99(2): 174–6.

CHAPTER 10: EXPRESS TRAIN TO HAPPINESS

1. Jean-Louis Millot, Gérard Brand, and Nadège Morand, "Effects of Ambient Odors on Reaction Time in Humans," *Neuroscience Letters* (April 5, 2002) 322(2): 79–82.

2. Reiko Sakamoto, Kazuya Minoura, Akira Usui, Yoshikazu Ishizuka, and Shigenobu Kanba, "Effectiveness of Aroma on Work Efficiency: Lavender Aroma During Recesses Prevents Deterioration of Work Performance," *Chemical Senses* (2005) 30(8): 683–91.

3. M. Luisa Demattè, Robert Österbauer, and Charles Spence, "Olfactory Cues Modulate Facial Attractiveness," *Chemical Senses* (2007) 32(6): 603–10.

4. Boris A. Stuck, Desislava Atanasova, Kathrin Frauke Grupp, and Michael Schredl, "The Impact of Olfactory Stimulation on Dreams." Paper presented at the 2008 Ameri-

can Academy of Otolaryngology–Head and Neck Surgery Foundation (AAO-HNSF) Annual Meeting & OTO EXPO in Chicago, September 21, 2008.

5. J. P. Royet, D. Zald, R. Versace, N. Costes, F. Lavenne, O. Koenig, and R.Gervais, "Emotional Responses to Pleasant and Unpleasant Olfactory, Visual, and Auditory Stimuli: A Positron Emission Tomography Study," *Journal of Neuroscience* (October 15, 2000) 20(20): 7752–9.

6. O. Pollatos, R. Kopietz, J. Linn, J. Albrecht, V. Sakar, A. Anzinger, R. Schandry, and M. Wiesmann, "Emotional Stimulation Alters Olfactory Sensitivity and Odor Judgment," *Chemical Senses* (July 2007) 32(6): 583–9.

7. J. P. Royet, J. Plailly, C. Delon-Martin, D. A. Kareken, and C. Segebarth, "fMRI of Emotional Responses to Odors: Influence of Hedonic Valence and Judgment, Handedness, and Gender," *Neuroimage* (October 2003) 20(2): 713–28.

8. C. Villemure, B. M. Slotnick, and M. C. Bushnell, "Effects of Odors on Pain Perception: Deciphering the Roles of Emotion and Attention," *Pain* (November 2003) 106(1–2): 101–8.

9. J. Lehrner, G. Marwinski, S. Lehr, P. Johren, and L.Deecke, "Ambient Odors of Orange and Lavender Reduce Anxiety and Improve Mood in a Dental Office," *Physiology and Behavior* (September 15, 2005) 86(1–2): 92–5.

10. Jeffrey J. Gedney, Toni L. Glover, and Roger B. Fillingim, "Sensory and Affective Pain Discrimination after Inhalation of Essential Oils," *Psychosomatic Medicine* (2004) 66: 599–606.

11. J. Willander and M. Larsson, "Smell Your Way Back to Childhood: Autobiographical Odor Memory," *Psychonomic Bulletin and Review* (April 2006) 13(2): 240–4.

12. J. Willander and M. Larsson, "Olfaction and Emotion: The Case of Autobiographical Memory," *Memory and Cognition* (October 2007) 35(7): 1659–63.

13. R. S. Herz, "A Naturalistic Analysis of Autobiographical Memories Triggered by Olfactory, Visual and Auditory Stimuli," *Chemical Senses* (March 2004) 29(3): 217–24.

14. Rachel Herz, *The Scent of Desire*. New York: William Morrow, 2007: 67.

15. R. S. Herz, J. Eliassen, S. Beland, and T. Souza, "Neuroimaging Evidence for the Emotional Potency of Odor-Evoked Memory," *Neuropsychologia* (2004) 42(3): 371–8.

16. G. N. Martin, "The Effect of Exposure to Odor on the Perception of Pain," *Psychosomatic Medicine* (July–August 2006) 68(4): 613–6.

17. S. Marchand and P. Arsenault, "Odors Modulate Pain Perception: A Gender-Specific Effect," *Physiology and Behavior* (June 1, 2002) 76(2): 251–6.

18. Xiuping Li, "The Effects of Appetitive Stimuli on Out-of-Domain Consumptive Impatience," *Journal of Consumer Research* (February 2008) 34: 649–56.

19. Janice K. Kiecolt-Glasera, Jennifer E. Graham, William B. Malarkey, Kyle Porter, Stanley Lemeshow, and Ronald Glaser, "Olfactory Influences on Mood and Autonomic, Endocrine, and Immune Function," *Psychoneuroendocrinology* (2008) 3: 328–39.

20. J. Djordjevic, J. N. Lundstrom, F. Clément, J. A. Boyle, S. Pouliot, and M.Jones-Gotman, "A Rose by Any Other Name: Would It Smell as Sweet?" *Journal of Neurophysiology* (January 2008) 99(1): 386–93.

21. Malarkey quote from "Aromatherapy May Make Your Feel Good, But It Won't Make You Well," Press release from Ohio State University, March 3, 2008, http://research news.osu.edu/archive/aromathe.htm. Accessed September 8, 2008.

CHAPTER 11: THE BITTER TRUTH

1. Linda Bartoshuk. "The Tongue Map," http://www.museumofhoaxes.com/hoax/weblog/permalink/the_tongue_map/. Accessed September 8, 2008.

2. L. A. Lucchina, O. F. Curtis, P. Putnam, A. Drewnowski, J. .M. Prutkin, and L. M. Bartoshuk, "Psychophysical Measurement of 6-n-propylthiouracil (PROP) Taste Perception," In C. Murphy (Ed.) *International Symposium on Olfaction and Taste XIX*. New York: New York Academy of Sciences (1998) 855: 816–9.

3. R. D. Mattes, "The Taste of Fat Elevates Postprandial Triacylglycerol," *Physiology and Behavior* (October 2001) 74(3): 343–8.

4. Michael G. Tordoff, "Chemosensation of Calcium," Symposium, "Chemical Senses and Health," Paper presented at the American Chemical Society National Meeting, Philadelphia, August 20, 2008.

5. Adam Drewnowski, Susan Ahlstrom Henderson, and Anne Barratt-Fornell, "Genetic Taste Markers and Food Preferences," *Drug Metabolism and Disposition* (2001) 29(4): 535–8.

6. Ibid.

7. Feng Zhang, Boris Klebansky, Richard M. Fine, Hong Xu, Alexey Pronin, Haitian Liu, Catherine Tachdjian, and Xiaodong Li, "Molecular Mechanism for the Umami Taste Synergism," *Proceedings of the National Academy of Sciences* (December 30, 2008): 105(52): 20930–4.

8. Jayaram Chandrashekar, Ken L. Mueller, Mark A. Hoon, Elliot Adler, Luxin Feng, Wei Guo, Charles S. Zuker, and Nicholas J. P. Ryba, "T2Rs Function as Bitter Taste Receptors," *Cell* (March 17, 2000) 100: 703–11; "Researchers Discover How Bitter Taste Is Perceived," Howard Hughes Medical Institute (March 10, 2005) http://www.hhmi.org/news/zuker7.html. Accessed January 5, 2008.

9. Ken L. Mueller, Mark A. Hoon, Isolde Erlenbach, Jayaram Chandrashekar, Charles S. Zuker, and Nicholas J. P. Ryba, " The Receptors and Coding Logic for Bitter Taste," *Nature* (March 10, 2005) 434: 225–9.

10. Y-Jen Huang, Yutaka Maruyama, Gennady Dvoryanchikov, Elizabeth Pereira, Nirupa Chaudhari, and Stephen D. Roper, "The Role of Pannexin 1 Hemichannels in ATP Release and Cell-Cell Communication in Mouse Taste Buds," *Proceedings of the National Academy of Sciences* (April 10, 2007): 104 (15): 6436–41.

11. G. K. W. Frank, T. A. Oberndorfer, A. N. Simmons, M. P. Paulus, J. L. Fudge, T. T. Yang, and W. H. Kaye, "Sucrose Activates Human Taste Pathways Differently from Artificial Sweetener," *NeuroImage* (February 15, 2008) 39(4): 1559–69.

12. I. E. de Araujo, A. J. Oliveira-Maia, T. D. Sotnikova, R. R. Gainetdinov, M. G. Caron, M. A. L. Nicholelis, and S. A. Simon, "Food Reward in the Absence of Taste Receptor Signaling," *Neuron* (March 27, 2008) 57: 930–41.

13. Fang-li Zhao, Tiansheng Shen, Namik Kaya, Shao-gang Lu, Yu Cao, and Scott Herness, "Expression, Physiological Action, and Coexpression patterns of Neuropeptide Y in Rat Taste-bud Cells," *Proceedings of the National Academy of Sciences* (August 2, 2005) 102 (31): 11100–5.

14. Quotes from Scott Herness. "Bitter or Sweet? The Same Taste Bud Can Tell the Difference," Press Release from Ohio State University, July 18, 2005, http://www.newswise.com/articles/view/513195. Accessed January 9, 2008.

15. Quote from Charles Zuker, "Researchers Discover How Bitter Taste is Perceived," Howard Hughes Medical Institution News Release, March 10, 2005, http://www.hhmi.org/news/zuker7.html. Accessed January 9, 2008.
16. Leslie B. Vosshall, "Putting Smell on the Map," *Trends in Neuroscience* (April 4, 2003) 4(2): 169–70.
17. M. L. Kringelbach, I. E. de Araujo, and E. T. Rolls, "Taste-Related Activity in the Human Dorsolateral Prefrontal Cortex," *Neuroimage* (February 2004) 21(2): 781–9.
18. J. O'Doherty, E. T. Rolls, S. Francis, R. Bowtell, and F. McGlone, "Representation of Pleasant and Aversive Tastes in the Human Brain," *Journal of Neurophysiology* (March 2001) 85(3) :1315–21.

CHAPTER 12: COCONUT CRAZY

1. Amy J. Tindell, Kyle S. Smith, Susana Pecina, Kent C. Berridge, and J. Wayne Aldridge, "Ventral Pallidum Firing Codes Hedonic Reward: When a Bad Taste Turns Good," *Journal of Neurophysiology* (2006) 96: 2399–409. See also Robert A. Wheeler and Regina M. Carelli. "The Neuroscience of Pleasure: Focus on 'Ventral Pallidum Firing Codes Hedonic Reward: When a Bad Taste Turns Good.'" *Journal of Neurophysiology* (2006) 96: 2175–6.
2. S. Fenu, V. Bassareo, and G. J. Di Chiara. "A Role for Dopamine D1 Receptors of the Nucleus Accumbens Shell in Conditioned Taste Aversion Learning," *Neuroscience* (September 1, 2001) 21(17):6897–904.

CHAPTER 13: COOKING UP SOME BRAIN CHEMISTRY

1. Jakob P. Ley, "Masking Bitter Taste by Molecules," *Chemosensory Perception* (2008) 1(1): 58–77.
2. Hervé This, *Kitchen Mysteries*. New York: Columbia University Press, 2007, 90–1.
3. Peter Barham, *The Science of Cooking*. Berlin: Springer-Verlag, 2001, 34.
4. H. Fukuda, S. Uchida, S. Kinomura, and R. Kawashima, "Neuroimaging of Taste Perception," *Advances in Neurological Sciences* (2004) 48(2): 285–93.
5. E. T. Rolls, "Smell, Taste, Texture, and Temperature Multimodal Representations in the Brain, and Their Relevance to the Control of Appetite," *Nutrition Review* (November 2004) 62(11 Pt 2): S193–204.
6. C. Viarouge, R. Caulliez, and S. Nicolaidis, "Umami Taste of Monosodium Glutamate Enhances the Thermic Effect of Food and Affects the Respiratory Quotient in the Rat," *Physiology and Behavior* (November 1992) 52(5): 879–84.
7. M. Jabbi, M. Swart, and C. Keysers, "Empathy for Positive and Negative Emotions in the Gustatory Cortex," *Neuroimage* (February 15, 2007) 34(4): 1744–53.
8. A Niijima, "Reflex Effects of Oral, Gastrointestinal and Hepatoportal Glutamate Sensors on Vagal Nerve Activity," *Journal of Nutrition* (April 2000) 130(4S Suppl): 971S–3S.
9. Nirupa Chaudhari, Ana Marie Landin, and Stephen D. Roper, "A Metabotropic Glutamate Receptor Variant Functions as a Taste Receptor," *Nature Neuroscience* (2000) 3: 113–9.

10. N.D Luscombe-Marsh, A. J. Smeets, and M. S. Westerterp-Plantenga, "Taste Sensitivity for Monosodium Glutamate and an Increased Liking of Dietary Protein," *British Journal of Nutrition* (April 2008) 99(4): 904–8.

CHAPTER 14: HOW TO KNIT A BABY TASTE BUD

1. Richard Restak , *The Secret Life of the Brain.* Washington, DC: Dana Press and Joseph Henry Press, 2001, 2–3.

2. Martin Witt and Klaus Reutter, "Embryonic and Early Fetal Development of Human Taste Buds: A Transmission Electron Microscopical Study," *The Anatomical Record* (December 1996) 246(4): 507–23; Martin Witt and Klaus Reutter. "Scanning Electron Microscopical Studies of Developing Gustatory Papillae in Humans," *Chemical Senses* (December 1997) 22(6): 601–12.

3. Fei Liu, Shoba Thirumangalathu, Natalie M. Gallant, Steven H. Yang, Cristi L. Stoick-Cooper, Seshamma T. Reddy, Thomas Andl, Makoto M. Taketo, Andrzej A. Dlugosz, Randall T. Moon, Linda A. Barlow, and Sarah E Millar, "Wnt-Beta-Catenin Signaling Initiates Taste Papilla Development," *Nature Genetics* (January 2007) 39(1): 106–1.

4. Ken Iwatsuki, Hong-Xiang Liu, Albert Gründer, Meredith A. Singer, Timothy F. Lane, Rudolf Grosschedl, Charlotte M. Mistretta, and Robert F. Margolskee, "Wnt Signaling Interacts with Shh to Regulate Taste Papilla Development," *Proceedings of the National Academy of Sciences* (February 13, 2007) 104(7): 2253–8.

5. C. M. Kotz, D. Weldon, C. J. Billington, and A.S. Levine, "Age-related Changes in Brain Prodynorphin Gene Expression in the Rat," *Neurobiology of Aging* (November, December 2004) 25(10): 1343–7.

CHAPTER 15: EXPECTING WHAT YOU TASTE AND TASTING WHAT YOU EXPECT

1. A. Soter, J. Kim, A. Jackman, I. Tourbier, A. Kaul, and R. L. Doty, "Accuracy of Self-Report in Detecting Taste Dysfunction," *Laryngoscope* (April 2008) 118(4): 611–7.

2. C. Strugnell, "Colour and Its Role in Sweetness Perceptions," *Appetite* (1997) 28: 85.

3. R. L. Alley and T. R. Alley, "The Influence of Physical States and Color on Perceived Sweetness," *Journal of Consumer Research* (1985) 12: 301–15.

4. Eric C. Koch, "Preconceptions of Taste Based on Color," *The Journal of Psychology* (May 2003) 137(3): 233–42.

5. R. A. Frank, K. Ducheny, and S. J. Mize, "Strawberry Odor, But Not Red Color, Enhances the Sweetness of Sucrose Solutions," *Chemical Senses* (1989) 14: 371–7.

6. Marcus Cunha Jr., Chris Janiszewski, and Juliano Laran, "Protection of Prior Learning in Complex Consumer Learning Environments." *Journal of Consumer Research* (April 2008) 34: 850–64.

7. W. K. Simmons, A. Martin, and L. W. Barsalou, "Pictures of Appetizing Foods Activate Gustatory Cortices for Taste and Reward," *Cerebral Cortex* (October 2005) 15(10): 1602–8.

8. S. Kikuchi, F. Kubota, K. Nisijima, S. Washiya, and S. Kato, "Cerebral Activation Focusing on Strong Tasting Food: A Functional Magnetic Resonance Imaging Study," *Neuroreport* (February 28, 2005) 16(3): 281–3.

9. C. Kobayashi, L. M. Kennedy, and B. P. Halpern, "Experience-Induced Changes in Taste Identification of Monosodium Glutamate (MSG) Are Reversible," *Chemical Senses* (May 2006) 31(4): 301–6.

10. Kristina M. Gonzalez, Catherine Peo, Todd Livdahl, and Linda M. Kennedy, "Experience Induced Changes in Sugar Taste Discrimination," *Chemical Senses* (February 1, 2008) 33(2): 173–9.

11. A. Faurion, B. Cerf, A. M. Pillias, and N Boireau, "Increased Taste Sensitivity by Familiarization to Novel Stimuli: Psychophysics, fMRI, and Electrophysiological Techniques Suggest Modulation at Peripheral and Central Levels." In C. Rouby (Ed.), *Olfaction, Taste and Cognition.* Cambridge: Cambridge University Press, 2002, 350–66.

12. Tom P. Heath, Jan K. Melicher, David J. Nutt, and Lucy P. Donaldson, "Human Taste Thresholds Are Modulated by Serotonin and Noradrenaline," *Journal of Neuroscience* (December 6, 2006) 26(49): 2664–71.

13. Ibid.

14. E. T. Rolls and C. McCabe. "Enhanced Affective Brain Representations of Chocolate in Cravers vs. Non-Cravers," *European Journal of Neuroscience* (August 26, 2007) 26(4):1067–76.

15. Jack B Nitschke, Gregory E. Dixon, Issidoros Sarinopoulos, Sarah J. Short, Jonathan D. Cohen, Edward E. Smith, Stephen M. Kosslyn, Robert M. Rose, and Richard J. Davidson, "Altering Expectancy Dampens Neural Response to Aversive Taste in Primary Taste Cortex," *Nature Neuroscience* (February 5, 2006) 9: 435–42.

16. F. Grabenhorst, E. T. Rolls, and A. Bilderbeck, "How Cognition Modulates Affective Responses to Taste and Flavor: Top-down Influences on the Orbitofrontal and Pregenual Cingulate Cortices," *Cerebral Cortex* (July 2008) 18(7): 1549–59.

17. Fabian Grabenhorst and Edmund T. Rolls, "Selective Attention to Affective Values Alters How the Brain Processes Taste Stimuli," *European Journal of Neuroscience* (2008) 27: 723–9.

CHAPTER 16: THE BIG PICTURE

1. The case studies are composites of multiple cases reported in the medical literature.

2. Helga Kolb, "How the Retina Works," *American Scientist* (January-February 2003) 91(1): 28.

3. Neuroscience for Kids, "The Retina," http://faculty.washington.edu/chudler/retina.html. Accessed July 3, 2008.

4. Ibid.

5. M. B. Manookin, D. L. Beaudoin, Z. R. Ernst, L. J. Flagel, and J. B. Demb, "Disinhibition Combines with Excitation to Extend the Operating Range of the OFF Visual Pathway in Daylight," *Journal of Neuroscience* (April 16, 2008) 28(16): 4136–50.

6. Z. M. Hafed and R. J. Krauzlis, "Goal Representations Dominate Superior Colliculus Activity During Extrafoveal Tracking," *Journal of Neuroscience* (September 17, 2008) 28(38): 9426–9.

7. Sayaka Sugiyama, Ariel A. Di Nardo, Shinichi Aizawa, Isao Matsuo, Michel Volovitch, Alain Prochiantz, and Takao K. Hensch, "Experience-Dependent Transfer of Otx2 Homeoprotein into the Visual Cortex Activates Postnatal Plasticity," *Cell* (August 8, 2008) 134: 508–20.

8. Quoted in "Trigger for Brain Plasticity Identified," Press Release from Children's Hospital Boston, August 7, 2008. http://www.eurekalert.org/pub_releases/2008–08/chb-tfb073108.php. Accessed August 15, 2008.

9. S. Zeki, "Thirty Years of a Very Special Visual Area, Area V5," *Journal of Physiology* (2004) 557 (1): 1–2.

10. From Nancy Kanwisher's research page, MIT, http://web.mit.edu/bcs/nklab/. Accessed July 3, 2008.

11. M. L. Kringelbach, A. Lehtonen, S. Squire, A. G. Harvey, M. G. Craske, I. E. Holliday, A. L. Green, T. Z. Aziz, P. C. Hansen, P. L. Cornelissen, and A. Stein, "A Specific and Rapid Neural Signature for Parental Instinct," *PloS ONE* (February 27, 2008) 3(2): e1664.

12. Neuroscience for Kids, "Brain Facts and Figures," http://faculty.washington.edu/chudler/facts.html. Accessed September 8, 2008.

13. H. Song, C. F. Stevens, and F. H. Gage, "Astroglia Induce Neurogenesis from Adult Neural Stem Cells," *Nature* (May 2, 2002), 39–44.

14. James Schummers, Hongbo Yu, and Mriganka Sur, "Tuned Responses of Astrocyctes and Their Influence on Hemodynamic Signals in the Visual Cortex," *Science* (June 20, 2008) 320: 1638–43.

15. Tzvi Ganel, Michal Tanzer, and Melvyn A. Goodale, "A Double Dissociation Between Action and Perception in the Context of Visual Illusions: Opposite Effects of Real and Illusory Size," *Psychological Science* (March 2008) 19(3): 221–5.

16. Lotfi B. Merabet, Roy Hamilton, Gottfried Schlaug, Jascha D. Swisher, Elaine T. Kiriakopoulos, Naomi B. Pitskel, Thomas Kauffman, and Alvaro Pascual-Leone, "Rapid and Reversible Recruitment of Early Visual Cortex for Touch," *PLOS One* (August 27, 2008). http://www.plosone.org/article/info%3Adoi%2F10.1371%2Fjournal.pone.0003046. Accessed September 6, 2008.

17. Quotes from "Scientists Unmask Brain's Hidden Potential," Press Release from Beth Israel Deaconess Medical Center," August 26, 2008. http://www.newswise.com/p/articles/view/543770. Accessed September 6, 2008.

18. Diego A. Gutnisky and Valentin Dragoi, "Adaptive Coding of Visual Information in Neural Populations," *Nature* (March 2008) 452: 220–4.

CHAPTER 17: COLOR AND MEMORY

1. Felix A. Wichmann, Lindsay T. Sharpe, and Karl R. Gegenfurtner, "The Contributions of Color to Recognition Memory for Natural Scenes," *Journal of Experiment Psychology: Learning, Memory, and Cognition* (2002) 28(3): 509–20.

2. Samuel G. Solomon and Peter Lennie, "The Machinery of Colour Vision," *Nature* (April 2007) 8: 276–86.

3. Karl R. Gegenfurtner, "Cortical Mechanisms of Colour Vision," *Nature Reviews Neuroscience* (July 2003) 4: 563–72.

4. Ibid.

5. Ibid.

6. Solomon and Lennie, 2007.

7. D. K. Murphey, D. Yoshor, and M. S. Beauchamp, "Perception Matches Selectivity in the Human Anterior Color Center," *Current Biology* (February 12, 2008) 18(3): 216–20.

8. Ibid.

9. K.R. Gegenfurtner and D.C. Kiper, "Color Vision," *Annual Review of Neuroscience* 26 (2003): 181–206.

10. Gegenfurtner, 2003.

11. Diane Ackerman, *An Alchemy of Mind: The Marvel and Mystery of the Brain*. New York: Scribner, 2004, 91.

12. Robert M. Sapolsky, "Stressed-Out Memories," *Scientific American Mind* (2004), 18(8): 28–34.

13. Richard Restak, *The New Brain: How the Modern Age Is Rewiring Your Mind*. Emmaus, PA: Rodale, 2003, 143.

14. E. Pastalkova, P. Serrano, D. Pinkhasova, E. Wallace, A. A. Fenton, and T. C. Sacktor, "Storage of Spatial Information by the Maintenance Mechanism of LTP," *Science* (August 25, 2006) 313(5790): 1141–4.

15. Wichmann, Sharpe, and Gegenfurtner, 2002.

CHAPTER 18: SIZING THINGS UP

1. All quotes from Scott Murray are personal communications.

2. Scott O. Murray, Huseyin Boyaci, and Daniel Kersten, "The Representation of Perceived Angular Size in Human Primary Visual Cortex," *Nature Neuroscience* (March 2006) 9(3): 429–34.

3. Quoted in Melanie Moran, "'Mind's Eye' Influences Visual Perception," Vanderbilt Exploration: News and Features," July 1, 2008, http://www.vanderbilt.edu/exploration/stories/mindseye.html . Accessed July 13, 2008.

4. Ibid.

5. Joel Pearson, Colin W. G. Clifford, and Frank Tong, "The Functional Impact of Mental Imagery on Conscious Perception," *Current Biology* (July 8, 2008) 18: 982–6.

6. S. O. Murray, D. Kersten, B. A. Olshausen, P. Schrater, and D. L. Woods, "Shape Perception Reduces Activity in Human Primary Visual Cortex," *Proceedings of the National Academy of Sciences* (2002) 99(23): 15164–9.

CHAPTER 19: ON THE MOVE

1. R.A.'s case is discussed in detail in L. M. Vaina, N. Makris, D. Kennedy, and A. Cowey, "The Selective Impairment of the Perception of First-Order Motion by Unilateral Cortical Brain Damage," *Visual Neuroscience* (1998) 15: 333–48.

2. F.D.'s case is discussed in detail in L. M. Vaina and A. Cowey, "Impairment of the Perception of Second Order Motion But Not First Order Motion in a Patient with Unilateral Focal Brain Damage," *Proceedings of the Royal Society of London, Series B, Biological Sciences* (1996): 263: 1225–32.

3. G. Johansson, "Visual Perception of Biological Motion and a Model for its Analysis," *Perception and Psychophysics* (1973) 14: 201–11.

4. See biological motion at http://www.lifesci.sussex.ac.uk/home/George_Mather/Motion/BM.HTML . Accessed June 23, 1008.

5. Anna Brooks, Ben Schouten, Nikolaus F. Troje, Karl Verfaillie, Olaf Blanke, and Rick van der Zwan, "Correlated Changes in Perceptions of the Gender and Orientation of Ambiguous Biological Motion Figures," *Current Biology* (September 9, 2008) 18: R728–9.

6. Francesca Simion, Lucia Regolin, and Hermann Bulf, "A Predisposition for Biological Motion in the Newborn Baby," *Proceedings of the National Academy of Sciences* (January 3, 2008) 105(2): 809–13.

7. If you are having trouble visualizing this, there's a good movie at http://www.lifesci.sussex.ac.uk/home/George_Mather/Motion/SECORD.HTML. Accessed June 23, 2008.

8. Zoe Kourtzi and Nancy Kanwisher, "Activation in Human MT/MST by Static Images with Implied Motion," *Journal of Cognitive Neuroscience* (2000) 12(1): 48–55.

9. Anja Schlack and Thomas D. Albright, "Remembering Visual Motion: Neural Correlates of Associative Plasticity and Motion Recall in Cortical MT," *Neuron* (March 15, 2007) 53: 881–90.

10. L. M. Vaina and S. Soloviev, "First-Order and Second-Order Motion: Neurological Evidence for Neuroanatomically Distinct Systems," *Progress in Brain Research* (2004) 144: 197–212.

11. Krystel R. Huxlin, Tim Martin, Kristin Kelly, Meghan Riley, Deborah I. Friedman, W. Scott Burgin, and Mary Hayhoe, "Perceptual Relearning of Complex Visual Motion after V1 Damage in Humans," *Journal of Neuroscience* (April 1, 2009) 29: 3981–91.

12. L. M. Vaina, A. Cowey, and D. Kennedy, "Perception of First- and Second-Order Motion: Separable Neurological Mechanisms," *Human Brain Mapping* (1999) 7: 67–77.

CHAPTER 20: VISION AND THE VIDEO GAME

1. C. Shawn Green and Daphne Bavelier, "Action Video Game Modifies Visual Selective Attention," *Nature* (May 29, 2003) 423: 534–7.

2. Ibid.

3. C. S. Green and D. Bavelier, "Action-Video-Game Experience Alters the Spatial Resolution of Vision," *Psychological Science* (2007) 18(1): 88–94.

4. C. S. Green and D. Bavelier, "Enumeration Versus Multiple Object Tracking: The Case of Action Video Game Players," *Cognition* (2006) 101: 217–45.

5. C. Shawn Green and Daphne Bavelier, "Effect of Action Video Games on the Spatial Distribution of Visuospatial Attention," *Journal of Experimental Psychology* (2006) 32(6): 1465–78.

6. M. W. G. Dye, C. S. Green, and D. Bavelier, "Transfer of Learning with Action Video Games: Evidence from Speed and Accuracy Measures" (submitted).

7. Quote attributed to Walpole in Richard Boyle, 2000, "The Three Princes of Serendip: Part One" http://livingheritage.org/three_princes.htm. Accessed June 6, 2008.

8. J. C. Rosser Jr., P. J. Lynch, L. Cuddihy, D. A. Gentile, J. Klonsky, and R. Merrell, "The Impact of Video Games on Training Surgeons in the 21st Century," *Archives of Surgery* (February 2007) 142(2): 181–6.

CHAPTER 21: HEARING, LEFT AND RIGHT

1. Kathryn D. Breneman, William E. Brownell, and Richard D. Rabbitt. "Hair Cell Bundles: Flexoelectric Motors of the Inner Ear," *PloS One* (April 22, 2009) 4(4): e5201.

2. Eckart O. Altenmüller, "Music in Your Head," *Scientific American: Mind* (January 2004): 24–31.

3. P. Rama, A. Poremba, J. B. Sala, L. Yee, M. Malloy, M. Mishkin, and S. M. Courtney, "Dissociable Functional Cortical Topographies for Working Memory Maintenance of Voice Identity and Location," *Cerebral Cortex* (2004) 14: 768–80.

4. Ibid.

5. "One Ear Is Not Like the Other, Study Finds." September 13, 2004. Scientific American.com. http://www.sciam.com/article.cfm?id=one-ear-is-not-like-the-o. Accessed August 10, 2008.

6. F. L. King and D. Kimura, "Left-ear Superiority in Dichotic Perception of Vocal Nonverbal Sounds," *Canadian Journal of Psychology* (June 1972) 26(2): 111–6.

7. Y. S. Sininger and B. Cone-Wesson, "Asymmetric Cochlear Processing Mimics Hemispheric Specialization," *Science* (September 10, 2004) 305(5690): 1581.

8. Y. S, Sininger and B. Cone-Wesson, "Lateral Asymmetry in the ABR of Neonates: Evidence and Mechanisms," *Hearing Research* (February 2006) 212(1–2): 203–11.

9. Quote in "Left and Right Ears Not Created Equal as Newborns Process Sound, Finds UCLA/UA Research," Press release, UCLA. September 9, 2004. http://www.eurekalert.org/pub_releases/2004–09/uoc—lar090804.php. Accessed August 12, 2008.

CHAPTER 22: LISTENING AND LANGUAGE

1. In Gardner Dozois (Ed.), *The Year's Best Science Fiction: Fourteenth Annual Collection*. New York: St. Martin's, 1997: 1–58.

2. Ibid, 33.

3. M. Pena, A. Maki, D. Kovacic, G. Dehaene-Lambertz, H. Koisumi, F. Bouquet, and J. Mehler, "Sounds and Silence: An Optical Topography Study of Language Recognition at Birth," *Proceedings of the National Academy of Sciences* (2003) 100(20): 11702–5.

4. A. D. Friederici, "The Neural Basis of Language Development and Its Impairment," *Neuron* (2006) 52: 941–52.

5. G. Dehaene-Lambertz, "Cerebral Specialization for Speech and Non-Speech Stimuli in Infants," *Journal of Cognitive Neuroscience* (2000) 12: 449–60.

6. D. R. Mandel, P. W. Jusczyk, D. G. Nelson, "Does Sentential Prosody Help Infants Organize and Remember Speech Information?" *Cognition* (1994) 53(2): 155–80.

7. Quoted in Faith Hickman Brynie, "Newborn Brain May Be Wired for Speech," *BrainWork* (July–August 2008), 7–8.

8. Jayaganesh Swaminathan, Ananthanarayan Krishnan, and Jackson T. Gandour, "Pitch Encoding in Speech and Nonspeech Contexts in the Human Auditory Brainstem," *NeuroReport* (July 16, 2008) 19(11): 1163–7.

9. Quoted in "Linguist Tunes in to Pitch Processing in Brain," February 16, 2008, Press release from Purdue University, http://www.eurekalert.org/pub_releases/2008–02/pu-lti011808.php. Accessed August 17, 2008.

10. Ibid.

11. For original sources on all these comparisons, see G. Dehaene-Lambertz, L. Hertz-Pannier, and J. Dubois, "Nature and Nurture in Language Acquisition: Anatomical and Functional Brain-Imaging Studies in Infants," *Trends in Neuroscience* (2006) 29(7): 367–73.

12. J. H, Gilmore, W. Lin, M. W Prastawa, C. B. Looney, Y. S. Vetsa, R. C. Knickmeyer, D. D. Evans, J. K. Smith, R. M. Hamer, J. A. Lieberman, and G. Gerig, "Regional Gray Matter Growth, Sexual Dimorphism, and Cerebral Asymmetry in the Neonatal Brain," *Journal of Neuroscience* (February 7, 2007) 27(6): 1255–60.

13. E. R. Sowell, P. M. Thompson, D. Rex, D. Kornsand, K. D. Tessner, T. L. Jernigan, and A. W. Toga, "Mapping Sulcal Pattern Asymmetry and Local Cortical Surface Gray Matter Distribution in Vivo: Maturation in Perisylvian Cortices," *Cerebral Cortex* (January 2002) 12(1): 17–26.

14. J. Dubois, M. Benders, A Cachia, F. Lazeyras, R. Ha-Vinh Leuchter, S. V. Sizonenko, C. Borradori-Tolsa, J. F. Mangin, and P. S. Hüppi," Mapping the Early Cortical Folding Process in the Preterm Newborn Brain," *Cerebral Cortex* (June 2008) 18(6): 1444–54.

15. J. Dubois, L. Hertz-Pannier, A. Cachia, J. F. Mangin, D. Le Bihan, and G. Dehaene-Lambertz, "Structural Asymmetries in the Infant Language and Sensori-Motor Networks," *Cerebral Cortex* (June 2008). [Epub ahead of print http://cercor.oxford journals.org/cgi/content/abstract/bhn097]. Accessed August 18, 2008.

16. K. Emmorey, J. S. Allen, J. Bruss, N. Schenker, and H. Damasio, "A Morphometric Analysis of Auditory Brain Regions in Congenitally Deaf Adults," *Proceedings of the National Academy of Sciences* (2003) 100(17): 10049–54.

17. G. Dehaene-Lambertz, S. Dehaene, and L. Hertz-Pannier, "Functional Neuroimaging of Speech Perception in Infants," *Science* (2002) 298: 2013–5.

18. Pena et al., 2003.

19. Dehaene-Lambertz et al., 2002.

20. Ibid.

21. J. R. Binder, J. A. Frost, T. A. Hammeke, P. S. Bellgowan, J. A. Springer, J. N. Kaufman, and E. T. Possing, "Human Temporal Lobe Activation by Speech and Nonspeech Sounds," *Cerebral Cortex* (2000) 10(5): 512–28.

22. D. Perani, S. Dehaene, F. Grassi, L. Cohen, S. F. Cappa, E. Dupoux, F. Fazio, and J. Mehler, "Brain Processing of Native and Foreign Languages," *NeuroReport* (1996) 7(15–17): 2439–44.

23. G. Schalug, S. Marchina, and A. Norton, "From Singing to Speaking: Why Singing May Lead to Recovery of Expressive Language Function in Patient's with Broca's Aphasia," *Music Perception* (2008) 25(4): 315–23.

24. B. Bonakdarpour, A. Eftekharzadeh, and H. Ashayeri, "Melodic Intonation Therapy in Persian Aphasic Patients," *Aphasiology* (2003) 17(1): 75–95.

25. J. Tamplin, "A Pilot Study into the Effect of Vocal Exercises and Singing on Dysarthric Speech," *NeuroRehabilitation* (2008) 23(3): 207–16.

26. G. Dehaene-Lambertz, L. Hertz-Pannier, J. Dubois, S. Meriaux, A. Roche, M. Sigman, and S. Dehaene, "Functional Organization of Perisylvian Activation During Presentation of Sentences in Preverbal Infants," *Proceedings of the National Academy of Sciences* (2006) 103(38): 14240–5.

27. Quoted in Brynie, 2008.

28. Gregory Hickok, Ursula Bellugi, and Edward S. Klima, "Sign Language in the Brain," *Scientific American* (June 2001): 46–53. Updated version available at http://lcn.salk.edu/publications/2008/HIckok_Sci_Amer_SI_2002.pdf. Accessed August 15, 2008.

29. M. Kashima, K. Sawada, K. Fukunishi, H. Sakata-Haga, H. Tokado, and Y. Fukui, "Development of Cerebral Sulci and Gyri in Fetuses of Cynomolgus Monkeys (*Macaca fascicularis*). II. Gross Observation of the Medial Surface," *Brain Structure and Function* (August 2008) 212(6): 513–20.

30. M. D. Hauser and K. Andersson, "Left Hemisphere Dominance for Processing Vocalizations in Adult, But Not Infant, Rhesus Monkeys: Field Experiments," *Proceedings of the National Academy of Sciences* (1994) 91(9): 3946–8.

31. C. Cantalupo, D. L. Pilcher, and W. D. Hopkins, "Are Planum Temporale and Sylvian Fissure Asymmetries Directly Related? A MRI Study in Great Apes," *Neuropsychologia* (2003) 41(14): 1975–81.

CHAPTER 23: MY EARS ARE RINGING

1. Alan H. Lockwood, Richard J. Salvi, and Robert F. Burkard, "Tinnitus," *New England Journal of Medicine* (September 9, 2002) 347(12): 904–10.

2. According to the American Tinnitus Association (ATA).

3. Ibid.

4. A. H. Lockwood, R. J. Salvi, R. F. Burkard, P. J. Galantowicz, M. L. Coad, and D. S. Wack, "Neuroanatomy of Tinnitus," *Scandinavian Audiology, Supplementum* (1999) 51: 47–52.

5. A. H. Lockwood, D. S. Wack, R. F. Burkhard, M. L. Coad, S. A. Reyes, S. A. Arnold, and R. J. Salvi, "The Functional Anatomy of Gaze-Evoked Tinnitus and Sustain Lateral Gaze," *Neurology* (February 27, 2001) 56(4): 472–80.

6. R. J. Pinchoff, R. F. Burkard, R. J. Salvi, M. L. Coad, and A. H. Lockwood, "Modulations of Tinnitus by Voluntary Jaw Movements," *American Journal of Otology* (November 19, 1998) 19: 785–9.

7. S. E. Shore, S. Koehler, M. Oldakowski, L. F. Hughes, and S. Syed, "Dorsal Cochlear Nucleus Responses to Somatosensory Stimulation Are Enhanced After Noise-Induced Hearing Loss," *European Journal of Neuroscience* (January 2008) 27: 155–68.

8. Shore quoted in "Overactive Nerves May Account for 'Ringing in the Ears,'" *Medical News Today*, January 11, 2008, http://www.medicalnewstoday.com/articles/93655.php. Accessed September 12, 2008.

9. Lockwood, Salvi, and Burkard, 2002.

10. A. Hatanaka, Y. Ariizumi, and K. Kitamura, "Pros and Cons of Tinnitus Retraining Therapy," *Acta Otolaryngolica* (April 2008) 128(4): 365–8.

11. J. A. Henry, M. A. Schecter, T. L. Zaugg, S. Griest, P. J. Jastreboff, J. A. Vernon, C. Kaelin, M. B. Meikle, K. S. Lyons, and B. J. Stewart, "Clinical Trails to Compare

Tinnitus Masking and Tinnitus Retraining Therapy," *Acta Otolaryngologica, Supplement* (December 2006) 556: 64–9.

12. National Alliance on Mental Illness, "Transcranial Magnetic Stimulation (TMS)," http://www.nami.org/Content/ContentGroups/Helpline1/Transcranial_Magnetic_Stimulation_(rTMS).htm. Accessed May 5, 2008.

13. J. A. Henry, C. Loovis, M. Montero, C. Kaelin, K. A. Anselmi, R. Coombs, J. Hensley, and K. E. James, "Randomized Clinical Trail: Group Counseling on Tinnitus Retraining Therapy," *Journal of Rehabilitation Research and Development* (2007) 44(1): 21–32.

14. C. Plewnia, M. Reimond, A. Najib, B. Brehm, G. Reischi, S. K. Plontke, and C. Gerloff, "Dose-Dependent Attenuation of Auditory Phantom Perception (Tinnitus) by PET-Guided Repetitive Transcranial Magnetic Stimulation," *Human Brain Mapping* (March 2007) 28(3): 238–46.

15. T. Kleinjung, P. Eichhammer, B. Langguth, P. Jacob, J. Marienhagen, G. Hajak, S. R. Wolf, and J. Strutz, "Long-Term Effects of Repetitive Transcranial Magnetic Stimulation (rTMS) in Patients with Chronic Tinnitus," *Otolaryngology: Head and Neck Surgery* (April 2005) 132(4): 566–9.

16. M. Mennemeier, K. C. Chelette, J. Myhill, P. Taylor-Cooke, T. Bartel, W. Triggs, T. Kimbrell, and J. Dornhoffer, "Maintenance Repetitive Transcranial Magnetic Stimulation Can Inhibit the Return of Tinnitus," *Laryngoscope* (July 2008) 118 (7):1228–32.

17. T. Kleinjung, P. Eichhammer, M. Landgrebe, P. Sand, G. Hajak, T. Steffens, J. Strutz, and B. Langguth, "Combined Temporal and Prefrontal Transcranial Magnetic Stimulation for Tinnitus Treatment: A Pilot Study," *Otolaryngology: Head and Neck Surgery* (April 2008) 138(4): 497–501.

18. Mennemeier et al., 2008.

CHAPTER 24: MUSIC AND THE PLASTIC BRAIN

1. Elizabeth W. Marvin and Elissa L. Newport, "Statistical Learning in Language and Music: Absolute Pitch Without Labeling," Presented at the International Conference on Music Perception and Cognition in Sapporo, Japan, August 26, 2008.

2. S. Baharloo, S. K. Service, N. Risch, J. Gitschier, and N. B. Freimer, "Familial Aggregation of Absolute Pitch," *American Journal of Human Genetics* (2000) 67: 755–58.

3. Andrea Norton, personal communication.

4. G. Schlaug, L. Jancke, Y. Huang, J. F. Staiger, and Helmuth Steinmetz, "Increased Corpus Callosum Size in Musicians," *Neuropsychologia* (August 1995), 1047–1055.

5. Norton, personal communication.

6. Charlie is not a real patient. He is a composite compiled from numerous research reports of the behavior and neurocognitive development of children who have Williams syndrome.

7. Allan L. Reiss, Stephan Eliez, J. Eric Schmitt, Erica Straus, Zona Lai, Wendy Jones, and Ursula Bellugi, "Neuroanatomy of Williams Syndrome: A High-Resolution MRI Study," *Journal of Cognitive Neuroscience* (2000) 12: S65–S73.

8. Albert M. Galaburda and Ursula Bellugi, "Multi-Level Analysis of Cortical Neuroanatomy in Williams Syndrome," *Journal of Cognitive Neuroscience* (2000) 12: S74–S88.

9. Ibid.

10. Mark A. Eckert, Albert M. Galaburda, Asya Karchemskiy, Alyssa Liang, Paul Thompson, Rebecca A. Dutton, Agatha D. Lee, Ursula Bellugi, Julie R. Korenberg, Debra Mills, Fredric E. Rose, and Allan L. Reiss, "Anomalous Sylvian Fissure Morphology in Williams Syndrome," *NeuroImage* (October 15, 2006) 3(1): 39–45.

11. Ibid.

12. Daniel J. Levitin, Vinod Menonc, J. Eric Schmitt, Stephan Eliez, Christopher D. White, Gary H. Glover, Jay Kadise, Julie R. Korenberg, Ursula Bellugi, and Allan L. Reiss, "Neural Correlates of Auditory Perception in Williams Syndrome: An fMRI Study," *NeuroImage* (January 2003) 18(1): 74–82.

13. M. Lotze, G. Scheler, H. R. Tan, C. Braun, and N. Birbaumer, "The Musician's Brain: Functional Imaging of Amateurs and Professionals during Performance and Imagery," *NeuroImage* (2003) 20(3): 1817–29.

14. Christian Gaser and Gottfried Schlaug, "Behavioral/Systems/Cognitive Brain Structures Differ Between Musicians and Non-Musicians," *Journal of Neuroscience* (October 8, 2003) 23(27): 9240–5.

15. F. H. Rauscher and M. A. Zupan, "Classroom Keyboard Instruction Improves Kindergarten Children's Spatial–Temporal Performance: A Field Experiment," *Early Childhood Research Quarterly* (2000) 15: 215–28.

16. L. K. Miller and G. I. Orsmond, "Cognitive, Musical and Environmental Correlates of Early Music Instruction," *Psychology of Music* (1999) 27(1): 18–37.

17. A. S. Chan, Y. C. Ho, and M. C. Cheung, "Music Training Improves Verbal Memory," *Nature* (1998) 396: 128.

18. Y. C. Ho, M. C. Cheung, and A. S. Chan, "Music Training Improves Verbal But Not Visual Memory: Cross-sectional and Longitudinal Explorations in Children," *Neuropsychology* (2003) 17(3): 439–50.

19. E. G. Schellenberg, "Music Lessons Enhance IQ," *Psychological Science* (2004) 15(8): 511–4.

20. K. Overy, "Dyslexia and Music: From Timing Deficits to Musical Intervention," *Annals of the New York Academy of Sciences* (2003) 999: 497–505.

21. Andrea Norton, Ellen Winner, Karl Cronin, Katie Overy, Dennis J. Lee, and Gottfried Schlaug, "Are There Pre-existing Neural, Cognitive, or Motoric Markers for Musical Ability?" *Brain and Cognition* (2005) 59: 124–34.

22. C. Gibson, B. S. Folley, and S. Park. "Enhanced Divergent Thinking and Creativity in Musicians: A Behavioral and Near-infrared Spectroscopy Study," *Brain and Cognition* (February 2009): 69(1): 162–9.

CHAPTER 25: COCHLEAR IMPLANT: ONE MAN'S JOURNEY

1. Stephen Michael Verigood, personal communication.

2. A. L. Giraud, C. J. Price, J. M. Graham, and R. S. Frackowiak, "Functional Plasticity of Language-Related Brain Areas After Cochlear Implantation," *Brain* (July 2001) 124(Pt 7):1307–16.

CHAPTER 26: SYNESTHESIA: WHEN SENSES OVERLAP

1. "Musical Taste: Synesthesia," *The Economist (US)* (March 5, 2005) 374 (8416): 79.

2. All quotes from Karyn Siegel-Maier are personal communications.

3. Richard Cytowic, *The Man Who Tasted Shapes*. New York: Putnam, 1993, 4.

4. Alison Motluk, "How Many People Hear in Colour?" *New Scientist* (May 27, 1995) 146: 18.

5. Judy Siegel-Itzkovich, "Listen to Those Blues and Reds," *Jerusalem Post* (January 7, 1996).

6. Julia Simner and Jamie Ward, "Synaesthesia, Color Terms, and Color Space: Color Claims Came from Color Names in Beeli, Esslen, and Jäncke (2007)," *Psychological Science* (April 2008) 19(4): 412–4.

7. R. J. Stevenseon and C. Tomiczek, "Olfactory-Induced Synesthesias: A Review and Model," *Psychological Bulletin* (2007) 133(2): 294–309.

8. J. Mishra, A. Martinez, T. Sejnowski, and S. A. Hillyard, "Early Cross-Modal Interactions in Auditory and Visual Cortex Underlie a Sound-Induced Visual Illusion," *Journal of Neuroscience* (April 11, 2007) 27(15): 4120–31.

9. J. D. Ryan, S. N. Moses, M. Ostreicher, T. Bardouille, A. T. Herdman, L. Riggs, and E. Tulving, "Seeing Sounds and Hearing Sights: The Influence of Prior Learning on Current Perception," *Journal of Cognitive Neuroscience* (January 22, 2008) 20: 1030–42.

10. B. E. Stein and M. A. Meredith, *The Merging of the Senses*. Cambridge, MA: MIT Press, 1993, 27.

11. Emiliano Macaluso and Jon Driver, "Multisensory Spatial Interactions: A Window onto Functional Integration in the Human Brain," *Trends in Neurosciences* (May 2005) 28(5): 264–71.

12. G. W. Septimus Piesse, *The Art of Perfumery*. Originally published in 1867. University of Michigan Historical Reprint Series, www.lib.umich.edu, page 85.

13. Burr, *The Emperor of Scent*, 109.

14. R. A. Osterbauer, P. M. Matthews, M. Jenkinson, C. F. Beckmann, P. C. Hansen, and G. A. Calvert, "Color of Scents: Chromatic Stimuli Modulate Odor Responses in the Human Brain," *Journal of Neurophysiology* (June 2005) 93(6): 3434–41.

15. S. Guest, C. Catmur, D. Lloyd, and C. Spence, "Audiotactile Interactions in Roughness Perception," *Experimental Brain Research* (September 2002) 146(2): 161–71.

16. M. L. Demattè, D, Sanabria, R. Sugarman, and C. Spence, "Cross-Model Interactions Between Olfaction and Touch," *Chemical Senses* (February 1, 2006) 31: 291–300.

17. G. Morrot, F. Brochet, and D. Dubourdieu, "The Color of Odors," *Brain and Language* (November 2001) 79(2): 309–20.

CHAPTER 27: IF THIS LOOKS LIKE DEJA VU, YOU AIN'T SEEN NUTHIN' YET

1. D. G. Rossetti, "Sudden Light" in *The Collected Works of Dante Gabriel Rossetti*. London: Ellis and Elvey: 1897, 295.

2. Takashi Kusumi, "Understanding the Déjà vu Experience," Chapter 14 in Kazuo Fujita and Shoji Itakura (Eds.), *Diversity of Cognition: Evolution, Development, Domestication and Pathology*. Balwyn North, Victoria, Australia: Trans Pacific Press, 2006, 310.

3. Louis Couperus, *Ecstasy: A Study of Happiness*. London: Henry and Company, 1897:16–17. Quoted in Herman N. Sno, Don H. Linszen, and Frans de Jonghe, "Art Imitates Life: *Déjà vu* Experiences in Prose and Poetry," *British Journal of Psychiatry* (1992) 160: 511–8.

4. I. A. Goncharov, *Oblomov*. New York: Penguin Books: 1980 [reprint from 1859], 470. Quoted in Sno et al., 514–5.

5. A. R. O'Connor and C. J. A. Moulin, "Normal Patterns of Déjà Experience in a Healthy, Blind Male: Challenging Optical Pathway Delay Theory," *Brain and Cognition* (December 2006) 62(3): 246–9.

6. T. Adachi, N. Adachi, Y. Takekawa, N. Akanuma, M. Ito, R. Matsubara, H. Ikeda, M. Kimura, and H. Arai, "Déjà Vu Experiences in Patients with Schizophrenia," *Comprehensive Psychiatry* (September–October 2006) 47(5): 389–93.

7. Edward Wild, "Déjà Vu in Neurology," *Journal of Neurology* (2005) 252: 1–7.

8. Adachi et al., 2006.

9. J. Bancaud, F. Brunet-Bourgin, P. Chauvel, and E.Halgren, "Anatomical Origin of Déjà Vu and Vivid 'Memories' in Human Temporal Lobe Epilepsy," *Brain* (February 1994) 117 (Pt 1): 71–90.

10. R. G. Thompson, C. J. A. Moulin, M. A. Conway, and R. W. Jones, "Persistent Déjà Vu: A Disorder of Memory," *International Journal of Geriatric Psychiatry* (2004) 19: 906–7.

11. "Jamais vu: Frequently Asked Questions," http://chris.moulin.googlepages.com/jamaisvufaq. Accessed April 6, 2008.

12. Alan S. Brown, *The Déjà Vu Experience: Essays in Cognitive Psychology*. New York: Psychology Press, 2004.

13. Alan S. Brown, "A Review of the Déjà Vu Experience," *Psychological Bulletin* (2003) 129(3): 394–413.

14. Alan S. Brown and Elizabeth J. Marsh, "Evoking False Beliefs About Autobiographical Experience," *Psychonomic Bulletin & Review* (2008) 15(1): 186–90.

15. Sergey Kalachikov, Oleg Evgrafov, Barbara Ross, Melodie Winawer, Christie Barker-Cummings, Filippo Martinelli Boneschi, Chang Choi, Pavel Morozov, Kamna Das, Elita Teplitskaya, Andrew Yu, Eftihia Cayanis, Graciela Penchaszadeh, Andreas H. Kottmann, Timothy A. Pedley, W. Allen Hauser, Ruth Ottman, and T. Conrad Gilliam, "Mutations in LGI1 Cause Autosomal-Dominant Partial Epilepsy with Auditory Features," *Nature Genetics* (2002) 30: 335–41.

16. Sandra Furlan, Federico Roncaroli, Francesca Forner, Libero Vitiello, Elisa Calabria, Salomé Piquer-Sirerol, Giorgio Valle, Jordi Perez-Tur, Roberto Michelucci, and Carlo Nobile, "The *LGI1/Epitempin* Gene Encodes Two Protein Isoforms Differentially Expressed in Human Brain," *Journal of Neurochemistry* (2006) 98(3): 985–91.

17. R. Michelucci, J. J. Poza, V. Sofia, M. R. de Feo, S. Binelli, F. Bisulli, E. Scudellaro, B. Simionati, R. Zimbello, G. d'Orsi, D. Passarelli, P. Avoni, G. Avanzini, P. Tinuper, R. Biondi, G. Valle, V.F. Mautner, U. Stephani, C. A. Tassinari, N. K. Moschonas, R. Siebert, A. L. de Munain, J. Perez-Tur, and C. Nobile, "Autosomal Dominant Lateral Temporal Epilepsy: Clinical Spectrum, New Epitempin Mutations, and Genetic Heterogeneity in Seven European Families," *Epilepsia* (October 2003) 44(10): 1289–97.

18. Wild, 2005.

19. F. Bartolomei, E. Barbeau, M. Gavaret, M. Guye, A. McGonigal, J. Régis, and P. Chauvel, "Cortical Stimulation Study of the Role of Rhinal Cortex in Déjà Vu and Reminiscence of Memories," *Neurology* (September 14, 2004) 63(5): 858–64.

20. Akira O'Connor, Amanda J. Barnier, and Rochelle E. Cox, "Déjà Vu in the Laboratory: A Behavioral and Experiential Comparison of Posthypnotic Amnesia and Posthypnotic Familiarity," *International Journal of Clinical and Experimental Hypnosis* (October 2008) 56(4): 425–50.

CHAPTER 28: FEELING A PHANTOM

1. R. Melznack, "Phantom Limbs and the Concept of a Neuromatrix," *Trends in Neuroscience* (1990) 13: 88–92.

2. L. E. Pezzin, T. R. Dillingham, and E. J. MacKenzie, "Rehabilitation and the Long-Term Outcomes of Persons with Trauma-Related Amputations," *Archives of Physical Medicine and Rehabilitation* (2000) 81: 392–300.

3. V. S. Ramachandran and William Hirstein, "The Perception of Phantom Limbs: The D. O. Hebb Lecture." *Brain* (1998) 121: 1602–30.

4. All quotes from Emma Duerden are personal communications.

5. C. H. Wilder-Smith, L. T. Hill, and S. Laurent, "Postamputation Pain and Sensory Changes in Treatment-Naive Patients: Characteristics and Responses to Treatment with Tramadol, Amitriptyline, and Placebo," *Anesthesiology* (2005) 103: 619–28.

6. Steven A. King, "Exploring Phantom Limb Pain," *Psychiatric Times*. (April 1, 2006) 23(4). http://www.psychiatrictimes.com/display/article/10168/49106?pageNumber=2. Accessed August 2, 2008.

7. Ambroise Paré, *The Workes of That Famous Chirurgion Ambrose Parey* (1649), Trans. by T. Johnson. London: Clarke, 338.

8. Nicholas Wade, "The Legacy of Phantom Limbs," *Perception* (2003) 32: 517–24.

9. E. S. Haldane and G. R. T. Ross, *The Philosophical Works of Descartes*. New York: Dover, 1955, 189.

10. Quoted in Nicholas Wade, "The Legacy of Phantom Limbs," http://www.artbrain.org/phantomlimb/wade.html . Accessed August 2, 1008. Attributed to W. Porterfield, *A Treatise on the Eye, the Manner and Phaenomena of Vision, Volume* 2, Edinburgh and Balfour: 1759, 362–5.

11. S. Weir Mitchell, *Injuries of Nerves and Their Consequences*. Philadelphia: Lippincott, 1972, 348.

12. Mitchell, 349.

13. Mitchell, 350.

14. Allan A. Bailey and Frederick P. Moersch, "Phantom Limb," *Canadian Medical Association Journal* (July 1941): 37–42.

15. Brenda L. Chan, Richard Witt, Alexandra P. Charrow, Amanda Magee, and Robin Howard, "Mirror Therapy for Phantom Limb Pain," *New England Journal of Medicine* (November 22, 2007) 357(21): 2206–7.

CHAPTER 29: PROBABILITIES AND THE PARANORMAL

1. Charles Richet, *Thirty Years of Psychical Research: Being a Treatise on Metapsychics*, Stanley De Brath (Translator). London: W. Collins Sons, 1923.

2. For details of the life of Charles Richet, see Stewart Wolf, *Brain, Mind, and Medicine: Charles Richet and the Origins of Physiological Psychology*. New Brunswick: Transaction Publishers, 1993.

3. Richet (1923), 3–4.

4. Richet (1923), 10.

5. R. Wiseman, E. Greening, and M. Smith, "Belief in the Paranormal and Suggestion in the Séance Room," *British Journal of Psychology* (August 3002) 93(Pt. 3): 285–97.

6. R. Wiseman and E. Greening, "'It's Still Bending': Verbal Suggestion and Alleged Psychokinetic Ability," *British Journal of Psychology* (February 2005) 96(Pt. 1): 115–27.

7. S. T. Moulton and S. M. Kosslyn, "Using Neuroimaging to Resolve the Psi Debate," *Journal of Cognitive Neuroscience* (January 2008) 21(1): 182–93.

8. Numerous (often conflicting) studies are reviewed in Richard Wiseman and Caroline Watt, "Belief in Psychic Ability and the Misattribution Hypothesis: A Qualitative Review," *British Journal of Psychology* (2006) 97: 323–38.

9. D. Pizzagalli, D. Lehmann, L. Gianotti, T. Koenig, H. Tanaka, J. Wackermann, and P. Brugger, "Brain Electric Correlates of Strong Belief in Paranormal Phenomena: Intracerebral EEG Source and Regional Omega Complexity Analyses," *Psychiatry Research* (December 22, 2000) 100(3): 139–54.

10. P. Brugger and K. I. Taylor, "ESP: Extrasensory Perception or Effect of Subjective Probability?" *Journal of Consciousness Studies* (2003) 10: 221–46.

11. Richet, 9.

CHAPTER 30: TIME

1. C. P. Cannon, C. H. McCabe, P. H, Stone, M. Schactman, B. Thompson, P. Theroux, R. S. Gibson, T. Feldman, N. S. Kleiman, G. H. Tofler, J. E. Muller, B. R. Chaitman, and E. Braunwald, "Circadian Variation in the Onset of Unstable Angina and Non-Q Wave Acute Myocardial Infarction (The TIMI III Registry and TIMI IIIB)," *American Journal of Cardiology* (February 1, 1997) 79(3): 253–8.

2. N.N. Jarjour, "Circadian Variation in Allergen and Nonspecific Bronchial Responsiveness in Asthma," *Chronobiology International* (September 1999) 16(5): 631–9.

3. D Holzberg and U. Albrecht, "The Circadian Clock: A Manager of Biochemical Processes Within the Organism," *Journal of Neuroendocrinology* (April 2003) 15(4): 339–43.

4. Michael Hastings, "A Gut Feeling for Time," *Nature* (May 23, 2002) 417(6887): 391–2.

5. M. Y. Cheng, C. M. Bullock, C. Li, A. G. Lee, J. C. Bermak, J. Belluzzi, D. R. Weaver, F. M. Leslie, and Q. Y. Zhou, "Prokineticin 2 Transmits the Behavioural Circadian Rhythm of the Suprachiasmatic Nucleus," *Nature* (May 23, 2002) 417(6887): 405–10.

6. Karen Wright, "Times of Our Lives," *Scientific American* (September 2002): 58–65.

7. Michael Tri H. Do1, Shin H. Kang, Tian Xue, Haining Zhong1, Hsi-Wen Liao, Dwight E. Bergles, and King-Wai Yau, "Photon Capture and Signalling by Melanopsin Retinal Ganglion Cells," *Nature* (January, 15 2009) 457: 281–7.

8. D. M. Berson, F. A. Dunn, and M. Takao, "Phototransduction by Retinal Ganglion Cells that Set the Circadian Clock," *Science* (February 8, 2002) 295 (5557): 955–7.

9. J. A. Perez-Leon, E. J. Warren, C. N. Allen, D. W. Robinson, and R. Lane Brown, "Synaptic Inputs to Retinal Ganglion Cells that Set the Circadian Clock," *European Journal of Neuroscience* (August 2006) 24(4): 1117–23.

10. C. B. Saper, T. C. Chou, and T. E. Scammell, "The Sleep Switch: Hypothalamic Control of Sleep and Wakefulness," *Trends in Neuroscience* (December 2001) 24(12): 726–31.

11. W. Mignot, S. Taheri, and S. Nishino, "Sleeping with the Hypothalamus: Emerging Therapeutic Targets for Sleep Disorders," *Nature Neuroscience Supplement* (November 2002) 5 (Suppl.): 1071–5.

12. C.B. Saper et al., 2001.

13. "Sleep Master Chemical Found," BBC News (January 16, 2003). http://news.bbc.co.uk/1/health/2664697.stm . Accessed June 9, 2008.

14. F. C. Chang and M. R. Opp, "Corticotropin-Releasing Hormone (CRH) as a Regulator of Waking," *Neuroscience and Biobehavioral Reviews* (July 2001) 25(5): 445–53.

15. C. Stetson, M. P. Fiesta, and D. M. Eagleman, "Does Time Really Slow During a Frightening Event?" *PloS ONE* (December 12, 2007) 2(12): ea2p5.

16. K. L. Toh, C. R. Jones, Y. He, E. J. Eide, W. A. Hinz, D. M. Virshup, L. J. Ptácek, and Y. H. Fu, "An h*Per2* Phosphorylation Site Mutation in Familial Advanced Sleep Phase Syndrome," *Science* (February 9, 2001) 291(5506): 1040–3.

17. T. I. Morgenthaler, T. Lee-Chiong, C. Alessi, L. Friedman, R. N. Aurora, B. Boehlecke, T. Brown, A. L. Chesson Jr., V. Kapur, R. Maganti, J. Owens, J. Pancer, T. J. Swick, and R. Zak, "Practice Parameters for the Clinical Evaluation and Treatment of Circadian Rhythm Sleep Disorders. An American Academy of Sleep Medicine Report," *Sleep* (November 2007) 30(11): 1445–59.

18. L. C. Lack and H. R. Wright, "Clinical Management of Delayed Sleep Phase Disorder," *Behavioral Sleep Medicine* (2007) 5(1): 57–76.

APPENDIX: THE BRAIN AND THE NERVOUS SYSTEM—A PRIMER

1. Daniel J. Levitin, *This Is Your Brain on Music: The Science of a Human Obsession.* New York: Penguin (Plume Edition), 2007, 96.

2. Patric Hagmann, Leila Cammoun, Xavier Gigandet, Reto Meuli, Christopher J. Honey, Van J. Wedeen, Olaf Sporns, "Mapping the Structural Core of Human Cerebral Cortex," *PLoS Biology* (July 1, 2008) (6)7: e159. http://biology.plosjournals.org/perlserv/?request=get-document&doi=10.1371/journal.pbio.0060159. Accessed August 18, 2008.

3. Ibid.

4. Society for Neuroscience, *Brain Facts: A Primer on the Brain and Nervous System.* Washington, DC: Society for Neuroscience, 2005, 10.

5. B. Pakkenberg, D. Pelvig, L. Marner, M. J. Bundgaard, H. J. G. Gundersen, J. R. Myengaard, and L. Regeur, "Aging and the Human Neocortex," *Experimental Gerontology* (January 2003): 95–9.

6. Anne D. Novitt-Morena, *How Your Brain Works* Emoryville. CA: Ziff-Davis, 1995, 15.

7. Richard Restak, *The Secret Life of the Brain.* Washington, DC: Dana Press and Joseph Henry Press, 2001, 5

8. Estimate from "The Brain's Inner Workings," National Alliance for the Mental Ill/Ohio, National Institute of Mental Health Outreach, http://namiohio.org/BRAINforweb.ppt. Accessed August 15, 2008.

9. Taulant Bacaj, Maya Tevlin, Yun Lu, and Shai Shaham, "Glia Are Essential for Sensory Organ Function in C. elegans," *Science* (October 31, 2008): 322 (5902): 744–7.

10. H. Song, C. F. Stevens, and F. H. Gage, "Astroglia Induce Neurogenesis from Adult Neural Stem Cells," *Nature* (May 2, 2002) 417: 39–44.

11. Restak, 5.

12. A. Arenz, R.A. Silver, A.T. Schaefer, and T.W. Margrie, "The Contribution of Single Synapses to Sensory Representation in Vivo," *Science* (August 15, 2008) 321: 977–80.

13. Ibid.

Recommended Resources

BOOKS

Blakeslee, Sandra, and Matthew Blakeslee. *The Body Has a Mind of Its Own: How Body Maps in Your Brain Help You Do (Almost) Everything Better.* New York: Random House, 2007.

Brown, Alan S. *The Déjà Vu Experience: Essays in Cognitive Psychology.* New York: Psychology Press, 2004.

Cole, Jonathan. *Pride and a Daily Marathon.* Cambridge, MA: MIT Press, 1995.

Cytowic, Richard E. *Synesthesia: A Union of Senses.* Cambridge, MA: MIT Press, 2002.

Doidge, Norman. *The Brain That Changes Itself: Stories of Personal Triumph from the Frontiers of Brain Science.* New York: Penguin, 2007.

Herz, Rachel. *The Scent of Desire: Discovering Our Enigmatic Sense of Smell.* New York: HarperCollins, 2007.

Levitin, Daniel J. *This Is Your Brain on Music: The Science of a Human Obsession.* New York: Penguin, 2006.

Meyers, Morton A. *Happy Accidents: Serendipity in Modern Medical Breakthroughs.* New York: Arcade, 2007.

Pinker, Steven. *The Stuff of Thought: Language as a Window into Human Nature.* New York: Penguin, 2007.

Sacks, Oliver. *Musicophilia: Tales of Music and the Brain.* New York: Alfred A. Knopf, 2007.

Tammet, Daniel. *Born on a Blue Day.* New York: Free Press, 2007.

This, Hervé. *Molecular Gastronomy: Exploring the Science of Flavor.* New York: Columbia University Press, 2006.

van Campen, Cretien. *The Hidden Sense: Synesthesia in Art and Science.* Cambridge, MA: MIT Press, 2008.

Wilson, Frank R. *The Hand.* New York: Random House, 1998.

ARTICLES

Ballantyne, Coco. "Losing Scents." *Scientific American* (December 2007) 297(6): 27–28.

Begley, Sharon. "How the Brain Rewires Itself." *Time* (January 29, 2007): 72–79.

Cunningham, Aimee. "Smells Like the Real Thing." *Science News* (July 15, 2006) 170(3): 40–41.

Foer, Joshua. "The Psychology of Déjà Vu." *Discover* (September 9, 2005). http://discover magazine.com/2005/sep/psychology-of-deja-vu.

Moseley, Lorimer. "Phantom Limb Cure: Retraining the Brain." *Scientific American* (September 9, 2008). http://www.scientificamerican.com/article.cfm?id=phantom-limb-cure-retrain-brain.

Motluk, Alison. "Particles of Faith." *New Scientist* (January 28, 2006): 34–36.

Murphy, Kate. "New Therapies Fight Phantom Noises of Tinnitus." *New York Times* (April 1, 2008). http://www.nytimes.com/2008/04/01/health/research/01tinn.html?em&ex=1207195200&en=6f028091d95aad08&ei=5087%0A.

Ramachandran, Vilayanur S., and Diane Rogers-Ramachandran, "I See, But I Don't Know," *Scientific American Mind* (December 2008/January 2009) 19(6): 20–22.

Ratliff, Evan. "Déjà Vu, Again and Again." *New York Times* (July 2, 2006). http://www.nytimes .com/2006/07/02/magazine/02dejavu.html.

Scientific American: Special Edition: Secrets of the Senses. (2006) 16(3).

Scientific American Reports: Special Edition on Perception. (2008) 18(2).

Shreve, James. "The Mind Is What the Brain Does." *National Geographic* (March 2005): 2–31.

Shulman, Matthew. "When Music Becomes Medicine for the Brain." *U.S. News and World Report* (September 1–8, 2008) 145(5): 89–90.

Smith, David V., and Robert F. Margolskee. "Making Sense of Taste." *Scientific American* (March 2001): 32+.

Stewart-Williams Steve. "The Placebo Puzzle: Putting Together the Pieces." *Health Psychology* (March 2004) 23(2): 198–206.

Weinberger, Norman M. "Music and the Brain." *Scientific American* (November 2004): 88–95.

WEB SITES

http://thebrain.mcgill.ca/flash/index_a.html. "The Brain from Top to Bottom" at this site offers answers and explanations for beginning, intermediate, and advanced users. It's maintained by McGill University in Montreal.

http://www.ninds.nih.gov/disorders/brain_basics/know_your_brain.htm. This site provides information on the basics of brain structure and function.

http://www.hhmi.org/senses/. This page features an article entitled, "Seeing, Hearing, and Smelling the World."

http://science.education.nih.gov/supplements/nih2/addiction/activities/activities_toc.htm. This site discusses interactive activities on the brain and provides information on how neurons work and the effects of drugs on the brain.

http://science.education.nih.gov/supplements/nih3/hearing/activities/activities_toc.htm. This site discusses how your brain interprets what you hear.

www.synesthete.org/index.php. This site has battery of tests you can take to determine if you are a synesthete.

www.senseofsmell.org. This site is maintained by the Sense of Smell Institute, which is the research and education division of the Fragrance Foundation.

https://www.ata.org/index.php. This is the American Tinnitus Association Web site.

http://www.parapsych.org. This is the Web site of Parapsychological Association, Inc., which is the international professional organization of scientists and scholars engaged in the study of "psi" (or "psychic") experiences, such as telepathy, clairvoyance, remote viewing, psychokinesis, psychic healing, and precognition. The Web site offers links to online experiments, quizzes, and references related to paranormal research.

http://www.brainconnection.com/topics/?main=anat/auditory-phys. This site offers an excellent explanation of how the ears and brain localize sound; see "How We Hear" by Ashish Ranpura.

http://www.hhmi.org/bulletin. This is the Web site of the Howard Hughes Medical Institute. It provides readable and informative summaries of the latest research on a variety of topics in the *HHMI Bulletin*, which can be downloaded.

http://www.dana.org. This is the Web site of the Dana Foundation, which publishes books, pamphlets, and articles about the brain and nervous system, often related to the senses. All are available here.

http://www.sciencedaily.com and www.sciencenews.org. Search these Web sites for readable accounts of what's new in brain science.

Index